T0347764

"WELL, DOC, YOU'RE IN"

"WELL, DOC, YOU'RE IN"

Freeman Dyson's Journey through the Universe

EDITED BY DAVID KAISER

The MIT Press
Cambridge, Massachusetts
London, England

The MIT Press would like to thank the anonymous peer reviewers who provided comments on drafts of this book. The generous work of academic experts is essential for establishing the authority and quality of our publications. We acknowledge with gratitude the contributions of these otherwise uncredited readers.

This book was set in Scala Pro and Scala Sans by Westchester Publishing Services. Printed and bound in the United States of America.

Library of Congress Cataloging-in-Publication Data is available.

ISBN: 978-0-262-04734-0

10   9   8   7   6   5   4   3   2   1

CONTENTS

CONTENTS

INTRODUCTION

DAVID KAISER

"Here is a scientist who can really write," the physicist Hans Bethe observed in his review of Freeman Dyson's first book, *Disturbing the Universe*, in 1979.[1] Dyson, who died on February 28, 2020, at the age of ninety-six, was a mathematician and theoretical physicist by training, but he became best known to most people as a writer. His first book, a poignant autobiographical reflection, was a finalist for the National Book Award; he went on to publish ten more. For twenty-five years, he was a regular contributor to the *New York Review of Books*, writing on a dazzling range of authors and topics—from Daniel Kahneman to Michael Crichton, the history of the Galápagos to the concept of infinity. His last piece, on "the outcast genius" and astrophysicist Fritz Zwicky, appeared six weeks before his death.[2]

Even by physicists' standards, Dyson's thinking was strikingly unconstrained by the here and now. One moment he was delving into the esoterica of quantum theory, and the next, he was speculating about the logistics of alien civilizations. In the 1950s, he joined the team developing a new type of nuclear reactor, which included several novel safety features; soon after, he was designing a spacecraft propelled by nuclear bombs. (His plans were scuttled by the 1963 Limited Test Ban Treaty.) Many of his views were penetrating; a few—like his insistence, late in life, that fears about rising carbon dioxide levels were misplaced—remained stubbornly out of step with the scientific consensus. To the end, he was a mental adventurer, not so much iconoclastic as intellectually fearless and relentlessly curious.

I first met Dyson in January 2001. He greeted me at the Institute for Advanced Study in Princeton, New Jersey, where he had worked, by then, for nearly a half century. Though he was in his late seventies, Dyson was elfish and spry, bounding up stairs two at a time. I was there to interview him for a book I was writing about how physicists had learned to calculate subtle effects among elementary particles using a theory known as quantum electrodynamics (QED). Dyson had played a pivotal role in those developments during the late 1940s, consolidating and pushing forward the insights of three other physicists—Richard Feynman, Julian Schwinger, and Sin-Itiro Tomonaga. These three later shared the Nobel Prize for the work, but it was Dyson's contribution that made QED a workable theory.[3]

That January afternoon, Dyson graciously sat with me for a two-hour interview, fielding my questions about the origins of QED—how the theory had evolved in collaboration with students and colleagues. As we were wrapping up, I sheepishly asked about something more personal. I'd seen some of Dyson's correspondence quoted in a colleague's book: letters he had written to his parents and his sister during some of the most exciting periods of his work. Could I see them? Without hesitating, Dyson hopped out of his chair, pulled open some file cabinets, and produced several thick folders bulging with letters. Even more remarkable, he set me up with a photocopier and a spare key to his office so I could make copies of the entire collection.

The letters were spellbinding. Stretching back to the 1940s, the weekly missives conveyed the uncertainty and exhilaration of Dyson's early forays into scientific research, interwoven with pitch-perfect vignettes about the people around him, such as his mentor at Cornell University, Bethe, and the irrepressible Feynman. Usually Dyson had written the letters by hand, but sometimes he had used an inexpensive portable typewriter. The originals give an impression of his quick mind at work; often stray letters appear above or below a given line, the typewriter's strained mechanisms no match for the speed of Dyson's thinking.

Much as his later essays would do, the letters roamed over a remarkable range of topics. His parents in England had never visited the United States, so Dyson played amateur anthropologist, sharing observations about postwar life. Thomas Dewey, the Republican governor of New York and 1948 presidential candidate, struck Dyson as "very greasy" during a visit to Cornell's campus, while "the notorious Henry Wallace" delivered an electrifying speech to Dyson's fellow students and teachers. Roaming beyond Ithaca, Dyson hitchhiked to New York City with friends and took long bus rides up and down the East Coast, into the Midwest, and all the way out to California. He sent thoughtful observations of race relations in Chicago, St. Louis, and Ypsilanti, Michigan, and described the "endless succession of rich well-tended farms and rich ill-tended industrial cities" on a long ride back East.[4] Those early letters conveyed a thirst—a physical, palpable thirst—to explore.

EXPANDING HORIZONS

Soon after Dyson passed away, I began working with several colleagues to examine facets of his life and legacy. The chapters that follow unfold in rough chronological order, charting Dyson's formative years, his budding interests and curiosities, and his efforts across the natural sciences, technology, and public policy. Rather than constituting a biography per se—for which Dyson's autobiographical writings and Phillip Schewe's *Maverick Genius*, published in 2013, provide valuable material—the chapters explore Dyson's particular way of thinking about deep questions, from the nature of matter, to the special challenges and opportunities afforded by nuclear power, to the origins of life and the ultimate fate of the universe.[5]

Born in Berkshire, England, in December 1923, Dyson was a mathematical prodigy. As Amanda Gefter describes in chapter 1, "That Secret Club of Heretics and Rebels," Dyson survived boarding school, found respite in musty libraries, and went on to study mathematics at the University of Cambridge. A determined autodidact, he developed a lifelong suspicion of organized curricula; he began crafting his persona as a scientific rebel at an early age. His studies at Cambridge were interrupted by war. In chapter 2, "Calculation and Reckoning: Navigating Science, War, and Guilt," William Thomas explores how Dyson learned to apply his quantitative skills as an analyst in the Operational Research Section of the Royal Air Force Bomber Command during the Second World War. There, too, Dyson quickly became disillusioned with what he perceived to be an ineffective and bureaucratic organization, hidebound by hierarchy.

Back at Cambridge after the war, Dyson switched from mathematics to physics but quickly grew restless. The real excitement in the field, he thought, had shifted to the United States, and so, with the aid of a Commonwealth Fellowship from the British government, he set off to pursue graduate study at Cornell in 1947. In chapter 3, "The First Apprentice," I follow Dyson's extraordinary progress in a subject—quantum electrodynamics—that had stymied the world's most accomplished theoretical physicists for decades. Within the two years covered by his fellowship, Dyson managed to synthesize new work by Feynman, Schwinger, and Tomonaga and push well beyond each of their calculations. When a young physicist extolled Dyson's contributions at a physics conference in New York City in January 1949, Feynman leaned over and announced, "Well, Doc, you're in." The only catch: Dyson never bothered to complete his PhD. To the end of his life, he proudly remained "Mr. Dyson."

Dyson's skyrocketing success garnered a prestigious research fellowship at the University of Birmingham (back in Britain) and a professorship at Cornell before J. Robert Oppenheimer managed to lure him with a lifelong appointment at the Institute for Advanced Study, beginning in 1953. As Robbert Dijkgraaf explains in chapter 4, "A Frog among Birds: Dyson as a Mathematical Physicist," Dyson took advantage of his new position—now relieved of the responsibilities of teaching and mentoring graduate students—to hop, frog-like, among mathematical puzzles that caught his eye. He focused on recondite topics at the border between mathematics and physics, producing the first rigorous demonstration that matter, as described by quantum theory, really should be stable and that mathematical techniques first honed for the study of matter in bulk could shed light on a century-old conundrum in number theory. Buoyed by his own experiences working at the unstable boundary between fields, Dyson lobbied two influential directors of the Institute—Oppenheimer and Carl Kaysen—to bolster the spirit of intellectual experimentation in the face of encroaching specialization.

Soon after settling into his new position at the Institute, Dyson began to expand his horizons still further. In chapter 5, "Single Stage to Saturn: Project Orion, 1957–1965," Dyson's son George Dyson describes how his father began consulting on civilian nuclear engineering projects with the defense contractor General Atomic.[6] Freeman Dyson reported to his parents in August 1956 that he had become intrigued—indeed, "absorbingly interest[ed]"—in the details of uranium reactors. Before long, he and his small team had produced a novel reactor design, still known by the acronym TRIGA (for Training, Research, Isotopes, General Atomic), which included several significant safety features. Not content to think through energy production here on Earth, Dyson quickly shifted his attention to the stars, working through designs for a new kind of spacecraft that could be propelled by exploding nuclear bombs. Political jockeying among federal agencies—not to mention increased concerns among citizens around the world over nuclear fallout in the atmosphere—ultimately scuttled the project but never dampened Dyson's enthusiasm for unconventional approaches to space travel.

Dyson's consulting with General Atomic led quickly to still broader engagements with science, technology, and public policy. As Ann Finkbeiner describes in chapter 6, "Dyson, Warfare, and the Jasons," in the early 1960s Dyson began working with a series of organizations—some outside the US federal government, some inside, and others that blurred the distinction.[7] His longest affiliation was with the Jasons, an elite scientific advisory group formed soon after the surprise launch of *Sputnik* to consult for the US Department of Defense and related agencies. Often the Jasons were called on to review classified projects and offer "reality checks" on some general's latest enthusiasm. Although controversy occasionally flared around the group's activities—especially during the Vietnam War—the work also triggered Dyson's novel contributions to adaptive optics, now a cutting-edge technique in astronomy.

From his childhood, Dyson had been intrigued by questions about the origins of life, even as he delved more deeply into mathematics and physics. During the 1960s and 1970s, spurred by discussions with biochemist Leslie Orgel during one of Dyson's sabbaticals at the University of California in San Diego, Dyson began to focus more squarely on biological topics. In chapter 7, "A Warm Little Pond: Dyson and the Origins of Life," Ashutosh Jogalekar explains how Dyson approached the question as a mathematical physicist, emphasizing simple interactions that can yield complex phenomena. He developed a model in which metabolism—the processes by which organisms consume energy to support motion, reproduction, and even thought—undergirds replication. Though certainly a minority view at the time, more recent work has effectively synthesized the metabolism-first approach with biologists' long-standing focus on replication.

In those same years, Dyson began to puzzle through questions of energy flow on much grander scales. As Caleb Scharf describes in chapter 8, "The Cosmic Seer," during the 1960s and 1970s Dyson explored unconventional questions about astrophysics, cosmology, and possible futures for life, intelligence, and the universe itself. As early as 1960, Dyson posited that advanced extraterrestrial civilizations—presumably desperate for reliable and efficient sources of energy—could repurpose all the matter of a modest-sized planet into an array of solar panels surrounding a star—and hence the "Dyson sphere" was born. To Dyson, such an audacious proposal quickly seemed quaint; after all, it concerned merely solar system–sized phenomena. By the late 1970s, he had worked out the energetics required for some bare form of consciousness—presumably long since stripped of our carbon-based embodiment—to persist indefinitely, into the impossibly distant future of an expanding and cooling universe.

In chapter 9, "A Bouquet of Dyson," Jeremy Bernstein recounts his half-century friendship with Dyson, key moments of which were conveyed via letter. Much like Dyson's remarkable letters home to his parents and sister, his correspondence with Bernstein unfolded in a kind of quantum superposition: deep and surprising insights about mathematical theorems and physical theories sprinkled amid casually wise observations about obscure novels, faraway places, academic politics, and family life. In turn, Bernstein, a theoretical physicist who was a staff writer for the *New Yorker* magazine for decades, served as "midwife," encouraging Dyson's early forays as an essayist and connecting him with magazine editors.

In the coda, "Not the End," Esther Dyson reflects on lessons learned from growing up with her father. As she recently explained, Freeman Dyson was "one of the youngest people to die at the age of ninety-six";

Freeman Dyson surrounded by his six children and sixteen grandchildren at his ninetieth birthday celebration, Institute for Advanced Study, September 2013. Photograph by George Dyson.

he retained a childlike curiosity to the end.[8] He also delighted in his own children: Esther and George, born during Dyson's marriage to the mathematician Verena Huber, whom Dyson met soon after he first arrived at the Institute for Advanced Study; and Dorothy, Emily, Mia, and Rebecca, born during Dyson's marriage to Imme Dyson (née Jung). Much as in his own research, Dyson encouraged his children not to feel too constrained by other people's rules; they were to follow their imaginations without losing sight of their responsibilities.

Accompanying these chapters are two series of original illustrations by Laurent Taudin. I have admired Taudin's work for years, especially his illustrations of Albert Einstein and concepts from relativity that have enriched a series of books by my colleagues.[9] In the main series of new illustrations, Taudin uses the graphic element of polyhedra to capture the multifaceted nature of Dyson's efforts and contributions. Dyson endeavored to think about questions from many angles; his own work reflected insights and methods from intellectual traditions that rarely aligned in straightforward ways. For the second series of illustrations, Taudin invokes musical metaphors and sensations of sound. Dyson's father was an accomplished musician, and young Dyson pursued the violin. But the *idea* of playing the violin made a deeper impression. He later explained to an interviewer that he played with mathematical tools because they were "beautiful," much as a violinist plays for love of the instrument.[10]

CHANGING THE CLIMATE

Freeman Dyson, a restless, relentless thinker, delighted in puzzling through complicated questions. Often he approached familiar, intractable problems from a novel perspective, while at other times, he surprised colleagues by broaching new questions they had not yet paused to consider. A self-styled rebel, Dyson found special joy in prodding conventional wisdom, always on the lookout for a weak spot or an overlooked assumption lurking within the scientific community's consensus.

Among his many scientific interests, Dyson brought this searching curi-osity to his early, creative engagements with the topic of climate change. To the frustration of many colleagues, Dyson's stance later calcified into an outlier position that could no longer be reconciled with the mounting body of evidence in the field—a legacy that few colleagues are likely to celebrate. His late-in-life pronouncements, delivered with rhetorical flair, were often amplified by outright climate change deniers, who claimed Dyson as one of their own.[11] Yet the sound bites and political jockeying of recent years have overshadowed the longer arc of Dyson's thinking on the topic. If we step back from his recent headline-grabbing assertions, we can identify several characteristics of his general approach to hard problems.

Dyson's first encounter with the topic of climate change—including early evidence that humans' industrial-age burning of fossil fuels was driv-ing accelerated warming—likely came at a conference held at the Insti-tute for Advanced Study in October 1955, just two years after he joined the Institute's faculty. Oppenheimer, as the Institute's director, helped to convene an international panel of experts, bringing together special-ists in geophysics, meteorology, and oceanography with early pioneers in programmable electronic computing. The Institute's own John von Neu-mann was a central participant in the conference. For years he had led a group at the Institute, building a novel computer powered by tens of thousands of vacuum tubes. Most of their calculations pertained to top-secret hydrogen bomb designs, but they also honed their new tool with unclassified simulations of weather and climate.[12]

Participants at the 1955 conference scrutinized data that seemed to indicate a recent warming trend—temperature records from around the world, changing rates of glacial recession in the Arctic and Antarctic—and also identified open questions about the circulation of carbon diox-ide among the atmosphere, the surface layer of the oceans, and the deep ocean. Some suggested that even a slight warming of the oceans could release substantial amounts of carbon into the atmosphere by disrupting the delicate balance of carbon-catching plant life such as plankton. Much of the discussion centered on the computational challenges involved in modeling such complicated, intertwined geophysical and biochemical processes.[13]

Dyson delved into the topic himself during a sabbatical in 1975–1976 at the Institute for Energy Analysis, an offshoot of the Oak Ridge National Laboratory in Tennessee. As he emphasized in a paper he published during his studies there, a "huge literature" had by then accumulated on "effects of the increasing CO_2 on the climate of the earth." Despite increasingly emphatic warnings by various analysts about an "acute ecological disaster" that could be triggered by rising levels of carbon dioxide in the atmosphere, much of the worldwide economy still depended on burning fossil fuels—a situation that, Dyson rightly surmised, would be difficult to change quickly. So, in characteristic fashion, he suggested that "the time is ripe to ask a different question."[14]

He sought to identify distinct strategies for short-term versus long-term responses to the "threat of catastrophe caused by CO_2." For long-term strategy, he was unequivocal: it "must be to stop burning fossil fuels and convert our industry to renewable photosynthetic fuels, nuclear fuels, geothermal heat and direct solar-energy conversion"—a necessary but disruptive process that he estimated would require several decades to implement, and hence "might well be too slow to avoid a CO_2 disaster." His short-term strategy was characteristically Dysonian: coordinate the planting and cultivation of large volumes of plants and trees, all over the world, that could sequester atmospheric carbon. Such a "carbon bank," he wrote, would not provide a long-term solution but would rather serve as a "stop-gap measure," to "buy time" until a "permanent shift from reliance on fossil fuels to renewable photosynthetic (or nuclear) fuels can be completed."[15]

For the remainder of his short paper, he ran the numbers: given current and projected rates of carbon emissions into the atmosphere each year, measurable improvements in the carbon balance could be achieved if large-scale but (in principle) feasible areas of land on several continents were prioritized for carbon-remediating plant life. The necessary land area, he further calculated, would not displace vital agricultural acreage.

To speed the process, he suggested that a modest carbon tax be applied to "every burner of fossil fuel": at a mere 0.021 cent per kilowatt hour (about 0.1 cent in 2022 US dollars), the low rate could be "easily supportable" while also sufficient for the task at hand.[16]

"It is highly unlikely that the particular emergency program here proposed will ever be implemented," he noted in his paper. "My discussion of it is enormously oversimplified. The purpose of this paper is to begin a process of mental preparation which may enable us to have realistic plans if ever the danger of catastrophe from CO_2 accumulation becomes acute." Straightforward scaling arguments—a favorite tool of mathematical physicists—demonstrated to Dyson's satisfaction that "there seems to be no law of physics or of ecology that would prevent us from taking action to halt or reverse the growth of atmospheric CO_2 within a few years if this should become necessary."[17]

Dyson made his next foray on the topic in an invited lecture at the University of Oxford in 1990. He began by highlighting one of the few "reliable numbers" regarding carbon emissions and climate change: the decades-long measurements of atmospheric carbon dioxide by Charles Keeling, which showed unmistakable evidence of rising carbon levels.[18] Dyson's main concern was that Keeling's chart showed *less* carbon accumulation in the atmosphere than might be expected, based on the rates of deforestation and fossil-fuel burning. That imbalance drove Dyson to return to processes he had first begun thinking about during his visit at Oak Ridge, such as the rates at which topsoil, peat, other plant matter, and forests absorb carbon, as well as processes that had been highlighted as early as the 1955 Institute conference, including exchanges between the atmosphere, the oceans' surface layer, and the deep ocean. As Dyson emphasized, information about several of these critical elements of the planet's overall carbon budget remained highly uncertain—so much so that whether they were net absorbers or emitters of carbon in a given year could not be pinned down.[19]

Dyson considered these large uncertainties to be deeply troubling because they obscured the underlying physical mechanisms by which carbon dioxide circulated and these key quantities remained as uncertain in 1990 as they had been in 1975. And so he chose to focus his lecture on the carbon problem, on which he would say "things that contradict the accepted wisdom. Where I diverge from the experts, the disagreement is not a matter of fact but of emphasis. I do not say that the experts are giving us wrong answers. I say that they are frequently not asking the right questions."[20]

He told his listeners that he had grown frustrated with the management of a large research program on environmental impacts of carbon dioxide, directed by the US Department of Energy, during his time at Oak Ridge. To Dyson's eyes, the program managers focused too narrowly on climate-impact considerations of carbon dioxide, even though rising carbon dioxide levels could have severe shorter-term impacts on agricultural yields and natural ecosystems as well. Moreover, too much of the research budget, to Dyson's taste, was directed toward computer simulations rather than fine-grained empirical measurements that might help to shrink the large uncertainties regarding various natural processes. Of course, improved computer models and simulations were important, Dyson was quick to note; after all, he had witnessed the remarkable progress that von Neumann's team at the Institute had made back in the early days of numerical computation. But he argued that the government should also devote expanded resources to other groups of scientists, successors to Charles Keeling, who "put on winter clothes and try to keep instruments correctly calibrated outside in the mud and rain."[21]

Dyson shared another criticism with his Oxford audience. He was leery of top-down bureaucratic approaches to any thorny problem—hearkening back to his wartime days with the Royal Air Force Bomber Command. The challenges of climate change and environmental stewardship called instead, Dyson insisted, for a more participatory model, akin to the campaign by the World Health Organization to eradicate smallpox: "centrally financed but locally administered and executed," with targeted engagement to convince people all around the world that their own best interests would be served by planting more trees that could improve local environments while also absorbing more atmospheric carbon.[22]

These were indeed differences in emphasis with most climate experts rather than a dispute over facts. Even in the face of such scientific uncertainties, Dyson nonetheless called for action: "Action to preserve the biosphere need not wait on detailed understanding. Understanding is not a prerequisite for action, and action is not a substitute for understanding. While we act, the quest for a deeper understanding must continue."[23]

Dyson wrote little about climate change for the next decade.[24] When he took to the topic again, late in 2005, almost none of his scientific arguments had changed, but his tone certainly had. "All the fuss about global warming is grossly exaggerated," he declared at Boston University that autumn. "Here I am opposing the holy brotherhood of climate model experts and the crowd of deluded citizens who believe the numbers predicted by computer models." He paused to note that "there is no doubt that parts of the world are getting warmer" and that such warming "obviously" causes "problems." But he declared that the broader community's focus on climate change diverted "money and attention from other problems that are more urgent and more important, such as poverty and infectious disease and public education and public health, and the preservation of living creatures on land and in the oceans." In the same breath, he decried the crumbling infrastructure and human neglect that had so devastatingly exacerbated the damage—especially to poor communities—when Hurricane Katrina struck New Orleans, just a few weeks before his lecture.[25]

By that time several climate experts had already accumulated evidence that climate change was contributing to the very list of problems that Dyson chose to highlight, ranging from infectious diseases and other threats to public health, to the increasing frequency and severity of hurricanes. Rather than engage with the more recent research, Dyson read a new lesson in his own older arguments. He suggested that a souped-up version of his 1970s plan, accelerated by (as-yet-unknown) biotechnology, might—at some point during "the next fifty years"—enable scientists to breed special types of trees and plants that could be especially efficient at sequestering carbon.[26]

More broadly, he argued that discussions about climate change at that time were more about values than about facts. On one side stood the "naturalists" for whom "any gross human disruption of the natural environment is evil" and who held that "anything we do to improve upon Nature will only bring trouble." In contrast, Dyson spoke of a "humanist ethic," according to which "humans are an essential part of nature," and hence "humans have the right and the duty to reconstruct nature so that humans and biosphere can both survive and prosper." Planting his staff firmly in the humanist camp, Dyson offered a techno-optimist view of "a time not far distant when genetically engineered food crops and energy crops will bring wealth to poor people in tropical countries and incidentally give us tools to control the growth of carbon dioxide in the atmosphere." Ironically, as he took to dismissing environmentalism (at this late stage) as "a worldwide secular religion," in the end he placed his own faith in a bio-technological future sustained—at the time he was writing—by wishes.[27]

What might account for Dyson's shift over the years in how he described the challenge of climate change? One striking feature of his interventions is an inverse relationship between how enmeshed he was in the details of various measurements and models and the vehemence with which he articulated his contrarian position. Well into the early 1990s, his discussions hewed close to the latest data, models, and scientific arguments as he crafted a distinct but complementary series of questions about a complicated topic. After he had stopped working directly on the topic—and after the scientific consensus about climate change had crystallized even more concertedly—he embraced the role of heretic, "telling stories that challenge the prevailing dogmas of today." Physicist and Nobel laureate Steven Weinberg captured the phenomenon well, explaining to a journalist, "I have the sense that when consensus is forming like ice hardening on a lake, Dyson will do his best to chip at the ice."[28]

*　*　*

Upon hearing the news of Dyson's death, I dug out the old stack of photocopies of his letters that I had made in his office all those years earlier.

The immediacy of the 1940s typeface and the uneven, hurried lines captured something essential about his mind in motion. Well beyond his work on quantum theory—as his intellectual enthusiasms splayed out ever more broadly throughout his long career—his thoughts remained no more bounded or neatly boxed than those early, jaunty letters had been. He was a once-in-a-generation thinker, doggedly exploring the cosmos and our place within it.

NOTES

Portions of this chapter originally appeared in David Kaiser, "Freeman Dyson's letters offer another glimpse of genius," *New Yorker,* 5 March 2020, and are reproduced with permission.

1. Hans Bethe, "Freeman Dyson's view of his times," *Physics Today* 32 (1979): 51–52; see also Freeman Dyson, *Disturbing the Universe* (New York: Harper, 1979).

2. Freeman Dyson, "The power of morphological thinking," *New York Review of Books* (16 January 2020).

3. David Kaiser, *Drawing Theories Apart: The Dispersion of Feynman Diagrams in Postwar Physics* (Chicago: University of Chicago Press, 2005); see also Silvan S. Schweber, *QED and the Men Who Made It: Dyson, Feynman, Schwinger, and Tomonaga* (Princeton: Princeton University Press, 1994).

4. Freeman Dyson to his parents, 16 October 1947 ("very greasy"), 16 December 1947 ("notorious Henry Wallace"), and 14 September 1948 ("well-tended farms"), reprinted in Freeman Dyson, *Maker of Patterns: An Autobiography Through Letters* (New York: Norton, 2018), 58, 65, and 99–100, respectively.

5. For (auto)biographical material, see esp. Dyson, *Disturbing the Universe*; Dyson, *Weapons and Hope* (New York: Harper and Row, 1984); Dyson, *Maker of Patterns*; Schweber, *QED and the Men Who Made It*, chap. 9; Phillip Schewe, *Maverick Genius: The Pioneering Odyssey of Freeman Dyson* (New York: St. Martin's Press, 2013); and William Press and Ann Finkbeiner, "Freeman J. Dyson, 1923–2020," in *Biographical Memoirs of the National Academy of Sciences* (2021), available at http://www.nasonline.org/publications/biographical-memoirs/memoir-pdfs/dyson-freeman.pdf. Silvan Schweber conducted a detailed oral-history interview with Dyson, available at http://www.webofstories.com/play/freeman.dyson/1.

6. General Atomic was established in 1955 as a division of the larger firm General Dynamics. During the late 1960s, the General Atomic division was sold to Gulf Oil and later to Royal Dutch/Shell. In 1986 the company was purchased by private owners, who changed its name to General Atomics. See George Dyson, *Project Orion: The True Story of the Atomic Spaceship* (New York: Holt, 2002), 32, 274.

7. Cf. Benjamin Wilson, *Insiders and Outsiders: Nuclear Arms Control Experts in Cold War America* (PhD diss., MIT, 2014).

8. Esther Dyson, email to the author, 23 February 2021.

9. See Taudin's illustrations throughout Michel Janssen and Christoph Lehner, eds., *The Cambridge Companion to Einstein* (New York: Cambridge University Press, 2014); Hanoch Gutfreund and Jürgen Renn, *The Road to Relativity* (Princeton: Princeton University Press, 2015); *The Formative Years of Relativity* (Princeton: Princeton University Press, 2017); and *Einstein on Einstein: Autobiographical and Scientific Reflections* (Princeton: Princeton University Press, 2020).

10. Dyson quoted in Thomas Lin, "Freeman Dyson: A 'rebel' without a Ph.D.," *Quanta* (26 March 2014), https://www.quantamagazine.org/a-math-puzzle-worthy-of-freeman-dyson-20140326.

11. See, e.g., Naomi Oreskes and Erik Conway, *Merchants of Doubt: How a Handful of Scientists Obscured the Truth on Issues from Tobacco Smoke to Global Warming* (New York: Bloomsbury Press, 2010), 61. In 2014, Dyson explained in an interview: "I spend maybe 1 percent of my time on climate, and that's the only field in which I'm opposed to the majority. . . . What I'm convinced of is that we don't understand climate, and so that's sort of a neutral position. I'm not saying the majority is necessarily wrong. I'm saying that they don't understand what they're seeing. It will take a lot of very hard work before that question is settled, so I shall remain neutral until something very different happens." Dyson quoted in Lin, "Freeman Dyson: A 'rebel' without a Ph.D."

12. George Dyson, *Turing's Cathedral: The Origins of the Digital Universe* (New York: Pantheon, 2012).

13. See especially the discussion section in Richard L. Pfeffer, ed., *Dynamics of Climate: The Proceedings of a Conference on the Application of Numerical Integration Techniques to the Problem of the General Circulation, Held October 26–28, 1955* (New York: Pergamon, 1960), 131–136. My thanks to George Dyson for bringing this reference to my attention. See also Paul Edwards, *A Vast Machine: Computer Models, Climate Data, and the Politics of Global Warming* (Cambridge, MA: MIT Press, 2010).

14. Freeman J. Dyson, "Can we control the carbon dioxide in the atmosphere?," *Energy* 2 (1977): 287–288.

15. Dyson, "Can we control the carbon dioxide in the atmosphere?," 288, 290.

16. Dyson, "Can we control the carbon dioxide in the atmosphere?," 289–290.

17. Dyson, "Can we control the carbon dioxide in the atmosphere?," 288, 290.

18. On Keeling's now-iconic data, see Joshua Howe, *Behind the Curve: Science and the Politics of Global Warming* (Seattle: University of Washington Press, 2014).

19. Freeman J. Dyson, "Carbon dioxide in the atmosphere and the biosphere," reprinted in Dyson, *From Eros to Gaia* (New York: Pantheon, 1992), 130–148, on 131–133. Dyson originally delivered this talk as the Radcliffe Lecture at Green College, Oxford, on 11 October 1990. My thanks to George Dyson for providing a typescript of the original lecture.

20. Dyson, "Carbon dioxide in the atmosphere and the biosphere," 135.

21. Dyson, "Carbon dioxide in the atmosphere and the biosphere," 136.

22. Dyson, "Carbon dioxide in the atmosphere and the biosphere," 144.

23. Dyson, "Carbon dioxide in the atmosphere and the biosphere," 148.

24. In a 2003 essay, Dyson wrote simply that "everyone agrees that the increasing abundance of carbon dioxide" drives several climactic effects, including "changes in rainfall, cloudiness, wind strength, and temperature," as well as accelerating sea-level rise, "which could become catastrophic if it continues to accelerate." As he had emphasized since his 1975 sabbatical at Oak Ridge, he also discussed possible effects on biological processes such as agricultural yields, though whether such changes could be beneficial in the short term—since plants require carbon dioxide for photosynthesis—remained far from clear. Freeman J. Dyson, "What a world!," *The Scientist as Rebel* (New York: New York Review of Books, 2006), 53–66, on 58, 62; the essay originally appeared in *New York Review of Books* (15 May 2003).

25. Freeman J. Dyson, "Heretical thoughts about science and society," in Dyson, *A Many-Colored Glass: Reflections on the Place of Life in the Universe* (Charlottesville: University of Virginia Press, 2007), 43–59, on 46–47. The material in that chapter was originally delivered as the Frederick S. Pardee Distinguished Lecture at Boston University in October 2005.

26. Dyson, "Heretical thoughts about science and society," 49. See, e.g., the scientific literature reviewed in C. Drew Harvell et al., "Climate warming and disease

risks for terrestrial and marine biota," *Science* 296 (21 June 2002): 2158–2162; and Kerry Emanuel, "A century of scientific progress," in *Hurricane! Coping with Disaster*, ed. Robert Simpson (Washington, DC: American Geophysical Union, 2003), 177–216, esp. section 12. More recent research continues to scrutinize Dyson's original idea from the 1970s, which he had introduced as an oversimplified scheme to help jump-start more careful analysis. See e.g., Julia K. Green et al., "Large influence of soil moisture on long-term terrestrial carbon update," *Nature* 565 (January 2019): 476–479; Bonnie Waring et al., "Forests and decarbonization: Roles of natural and planted forests," *Frontiers in Forests and Global Change* 3 (May 2020), article 58; and Connor J. Nolan, Christopher B. Field, and Katharine J. Mach, "Constraints and enablers for increasing carbon storage in the terrestrial biosphere," *Nature Reviews Earth and Environment* 2 (May 2021): 436–446.

27. Dyson, "Heretical thoughts on science and society," 55–56. Dyson described environmentalism as "a worldwide secular religion" in "The question of global warming," in Dyson, *Dreams of Earth and Sky* (New York: New York Review of Books, 2015), 81–98, on 95. Dyson wrote of the promise of genetic engineering and biotechnology well beyond the topic of climate change in several of his books and essays. See especially Freeman J. Dyson, *Infinite in All Directions* (New York: Harper Perennial, 1989); Dyson, *Imagined Worlds* (Cambridge, MA: Harvard University Press, 1997); Dyson, *The Sun, The Genome, and the Internet: Tools of Scientific Revolutions* (New York: Oxford University Press, 1999); Dyson, *The Scientist as Rebel*; and Dyson, *A Many-Colored Glass*.

28. Steven Weinberg quoted in Nicholas Dawidoff, "The civil heretic," *New York Times* (25 March 2009). A helpful indication of widespread scientific consensus on climate change around the time that Dyson gave his outspoken lecture at Boston University is Naomi Oreskes, "The scientific consensus on climate change," *Science* 306 (3 December 2004): 1686.

1 THAT SECRET CLUB OF HERETICS AND REBELS

AMANDA GEFTER

In winter 1991, sixty-seven-year-old Freeman Dyson stepped on stage to receive the Oersted Medal. Dyson, a mathematician and physicist at the Institute for Advanced Study in Princeton, New Jersey, and a fellow of the Royal Society, was famous worldwide for his groundbreaking work in quantum electrodynamics, his design of a nuclear reactor, his plans for interstellar spacecraft, and his imaginings of alien civilizations, not to mention his beloved popular books. But the Oersted was a *teaching* award—a medal for having an "outstanding, widespread, and lasting impact on the teaching of physics." It was a prestigious honor; Carl Sagan had won the year before. Dyson thanked the American Association of Physics Teachers for the prize, then delivered his acceptance speech.

"I find it absurd that I should have been chosen to receive this honor," Dyson began. "The Oersted was one medal that I was sure I would never get. When I look around, among all my friends there is nobody who does less teaching than I do. There is nobody who deserves this medal less than I do."[1]

It wasn't just his lack of teaching experience that made Dyson uncomfortable accepting the prize. It was that he didn't really believe in the idea of teaching physics. The PhD system was bad enough—*Mr.* Dyson vehemently opposed it. But it was more than that. It was science education of all kinds that was the problem—especially for children.

"Science is presented to our young people as a rigid and authoritarian discipline," Dyson said to the crowd of educators. But science is meant to be a subversion of authority, an international secret club of heretics and rebels. Teachers, he said, are authority figures, and that makes teaching physics a contradiction in terms. "If science ceases to be a rebellion against authority," Dyson warned, "then it does not deserve the talents of our brightest kids."[2]

THE START OF AN INFINITE SERIES

Freeman John Dyson was one of those brightest kids. Born December 15, 1923, in Crowthorne, England, to George Dyson, a musician and composer who would later be knighted, and Mildred Atkey Dyson, a lawyer and social worker, Freeman's thoughts were dancing with numbers from the start. According to his own recollections, he was in his crib, and was supposed to be napping, when he tried adding one plus a half plus a quarter plus an eighth plus a sixteenth, and so on. It dawned on him that if he kept going he'd eventually end up with two. Then he tried one plus a third plus a ninth, and kept going, and saw the sequence heading toward one and a half.

"I had discovered infinite series," Dyson recalled. "I don't remember talking about this to anybody at the time. It was just a game I enjoyed playing."[3] At age four, he'd attempt to calculate the number of atoms in the sun. "Bright" doesn't do justice to Dyson's mathematical mind. The kid was a prodigy.

George and Mildred saw their son's precociousness, but they didn't foster it so much as stay out of the way. True upper-middle-class Brits, they were remote figures in their children's lives, leaving Freeman and his sister, Alice, to be raised by the staff: a nanny, a cook, a housekeeper, and a gardener. Once a week the children were allowed to eat dinner with their parents; otherwise they were to admire them from afar. Once, when Alice went so far as to hug her mother, she was scolded for messing up her mom's hair.[4] The Dysons didn't fill their home with warmth exactly, but they did fill it with culture, music, and—best of all—books. Oddly, though, they didn't always let the kids read them. When, for instance, George brought home a copy of E. T. Bell's *Men of Mathematics*, the book beckoned to Freeman like none had before, but George returned it to the library before Freeman had a chance to crack open the cover.[5] Freeman would never forget that.

Maybe that's why books took on a certain forbidden allure. Freeman scavenged his parents' bookshelves, devouring popular science books by Alfred North Whitehead, James Jeans, Lancelot Hogben, and J. B. S. Haldane. At age seven, he read Arthur Eddington's descriptions of relativity in *Space, Time and Gravitation*, where he gazed upon a diagram dividing space-time into two so-called light cones, one containing an observer's absolute past and the other his absolute future. Everything outside the bounds of the light cones was labeled "Elsewhere." When the nanny came into the room where Freeman was reading and asked him where his sister was, he, unsure, replied, "In the absolute Elsewhere." George, hearing the story from the nanny, found it so amusing that he wrote it up and sent it to *Punch* magazine, which published it in the form of a cartoon.

Freeman couldn't see why the cartoon was funny. But it *was* funny. Not haha funny, but prophetic funny. After all, in the light cone picture of relativity, "elsewhere" is a profoundly unreachable, unknowable, unplace—a region impenetrable not only in practice but in principle. Surely the seven-year-old genius understood that, and yet, in his offhand comment, Freeman referred to the "absolute Elsewhere" as a kind of epistemological hurdle, a place that, given a bit more information—more numbers, more books, more rocket fuel—he could surely conquer, a region containing things he simply didn't know—*yet*. He didn't read the "absolute" as a warning so much as a challenge and a dream.

Though they had some science books lying around, the Dyson family wasn't particularly science-minded. His mother loved literature; his father was a brilliant musician. The music director of Winchester College, a renowned boy's high school, George was known for giving wonderfully entertaining lectures on the history and theory of music, performing Bach or Brahms on the piano while simultaneously expounding on the meaning of the pieces he was playing. The lectures were so popular that in 1930, the BBC asked him to give a series of them in their studios, which they then aired. The producers were shocked when Dyson shunned a script—he preferred to wing it, playing and speaking off the cuff from the piano bench.[6]

Freeman wasn't untalented musically speaking; he played the violin well, but he played mathematics even better. "I had this skill with mathematical tools," he said, "and I played these tools as well as I could, just because it was beautiful, rather in the same way a musician plays a violin . . . just because he loves the instrument."[7]

"I played these [mathematical] tools as well as I could, just because it was beautiful, rather in the same way a musician plays a violin . . . just because he loves the instrument."

There was one Dyson who was scientifically minded: Sir Frank Dyson, the Astronomer Royal. Technically he wasn't related to Freeman, but as far as Freeman was concerned, that didn't matter. "We were proud to share his name," Freeman wrote. "His glory helped to turn my infant thoughts toward astronomy."

In summer 1927 there was a solar eclipse, which Freeman learned would be total at Giggleswick, a town two hundred miles north of Winchester. He was desperate to see it and became furious when George refused to take him there, telling Freeman he could catch the next one—in 1999.[8] Angrily, Freeman held bits of glass over a smoky candle to darken them, then bitterly watched the sun shrink to a narrow crescent, but no smaller.

He read works of science fiction, drew diagrams of the solar system, contemplated the possibilities of other forms of life. He penned his own work of science fiction—*Sir Phillip Roberts's Erolunar Collision*—and based the story's hero on Frank Dyson. Freemen grew increasingly convinced, as he put it, that "men would reach the planets in my lifetime, and that I should help in the enterprise."[9] The future light cone beckoned.

Freeman's dreams of the future made him something of an outsider in Winchester, the medieval cathedral town where they'd lived since shortly after he was born. Everyone there was planted firmly in the past. The Dysons' house was three hundred years old, everything around them even older. The ancient capital of England, Winchester had a long prehistory— most recently as the Roman town Venta Belgarum, established circa 70 CE, and as an Iron Age hill fort settlement before that, dating back to 150 BCE. The result of all that history was a present layered with the past—a maze of medieval houses, castles, and winding market streets built atop Roman temples, aqueducts, and graves, yesteryear's crumbling stone monasteries and garden walls peeking up through excavation sites like tips of icebergs. Iron Age jewelry, Roman coins, and fragments of Anglo-Saxon pottery were sedimented into the sidewalks beneath the Dysons' feet as they walked from their house on Kingsgate to the gates of Winchester College. There, while George was giving music lessons, Freeman and Alice ran through ancient water meadows, climbed ancient trees, fished for minnows and sticklebacks, and chased their dog among the six-hundred-year-old buildings on the campus of the oldest private school in England.[10]

In the town of Winchester, Dyson wrote, "The past is there, close and tangible. The people around me were constantly discussing the fine points of our local history, the details of medieval church architecture, or the latest discovery in the archaeological diggings that were all the time in progress. With the impatience of a child, I reacted strongly against all this. Why were these people so stuck in the past? Why were they so excited about some bishop who lived six hundred years ago? I did not want to go back six hundred years into that dull old world that they loved so much. I would much rather go six hundred years forward. So while they talked learnedly of Chaucer and William of Wykeham, I dreamed of spaceships and alien civilizations."[11] It was his first act of rebellion—his initiation into that international secret club.

SEEKING SOLACE IN A SCIENCE SOCIETY

"We all know that the decisive years for turning young minds away from science are the years before they reach high school," Dyson continued from the American Association of Physics Teachers' award stage. "The damage is mostly done in elementary and middle schools. Again, I use my own experience to show how apparently unfavorable circumstances may lead to favorable results. Here is what happened to me."

In 1932, the eight-year-old Dyson was shipped off to boarding school. "Shipped off" isn't quite the right term—the Twyford school was within walking distance of the Dysons' home. But for Freeman, that only made it more unjust. He was stuck there just like all the other boys, and his parents never came to visit.

Twyford was a brutal place—the kind where the headmaster carries a whip and eagerly uses it on boys who bungle their Latin grammar. Freeman wasn't the kind of boy to bungle grammar, Latin or otherwise, and so he mostly avoided the whip, but that only made things worse for him:

it put a target on his back. Acing his tests and winning prizes, Freeman skipped grade after grade after grade until, in a matter of months, the scrawny eight-year-old had landed himself in the Upper School, where his fellow classmates were suddenly much older and much bigger. Prodigy has its downsides.

At Twyford, that downside came in the form of sandpaper. "So far as I was concerned, the cruelty of the headmaster was less oppressive than the brutality of the boys," Dyson said. "The boys were at their most barbarous age, and they redressed the balance of injustice by torturing those who had escaped whipping. Their favorite instrument of torture, sandpaper applied to the face or to other tender areas of skin, was more to be feared than the headmaster's whip."[12]

Maybe it came in handy that Dyson, though small, could run. According to school records, he was accomplished in track and won competitions in hurdles and high jump with, as the *Twyfordian* put it, "the agility of *pulex irritans*"—a common flea.[13] But Dyson had a better escape plan than running; when the going got rough, he ducked into Twyford's singularly magical place: its library. "I could always escape to the library," he said. "The more barbaric [the school] was, the better the library seemed to be."[14]

There he found the works of Jules Verne and a children's encyclopedia, *Book of Knowledge*, wherein he read about electrons and electricity and found himself wondering about protons and "proticity."[15] Was there such a thing, he wondered? He wouldn't learn the answer in class—because Twyford didn't teach science. "The wave of [educational] reform, which by that time had begun to introduce serious science teaching into the high schools, had not yet penetrated down to the Prep Schools," Dyson said. "At the Prep School we were still taught the classical 19th-century curriculum, lots of Latin, a good deal of mathematics, no science."[16] And that, Dyson argued to the Oersted crowd, is exactly why he became a scientist.

"So it happened that I belonged to a small minority of boys who were lacking in physical strength and athletic prowess, interested in other things besides football, and squeezed between the twin oppressions of whip and sandpaper. We hated the headmaster with his Latin grammar and we hated even more the boys with their empty football heads. So what could the poor helpless minority of intellectuals, later and in another country to be known as nerds, do to defend ourselves? We found our refuge in a territory that was equally inaccessible to our Latin-obsessed headmaster and to our football-obsessed schoolmates. We found our refuge in science."

"With no help from the school authorities, we founded a Science Society," Dyson continued. "As a persecuted minority we kept a low profile. We held our meetings quietly and inconspicuously. We couldn't do any real experiments. All we could do was share books and explain to each other what we didn't understand. But we learned a lot. Above all, we learned those lessons that can never be taught by formal courses of instruction, that science is a conspiracy of brains against ignorance, that science is a revenge of victims against oppressors, that science is a territory of freedom and friendship in the midst of tyranny and hatred."[17]

Dyson left Twyford in 1936. He always looked back on those four hellish years as the worst of his life, a traumatic, lingering horror. He would remember it when it came time to send his own small children to school, which he would do in the United States, to public schools, as far as possible from the gothic cruelty of old British prep. He would remember it too when Twyford invited their famous alum to return to the school in 2009 to dedicate their Saxon Court, a new set of classrooms named after a set of medieval, Saxon-era graves discovered at the site; his wife would have to coax him to go and confront the dark corners of his childhood.[18] He did go, reluctantly, donning a necktie emblazoned with the formula $E = mc^2$, a tie given to him by Albert Einstein, member of that international secret club of heretics and rebels.[19]

"Perhaps my experience in that Prep School between the ages of eight and 12 was not exceptional," Dyson said from the Oersted stage. "Perhaps there were others in similar circumstances who found in science a beacon of freedom and hope. Perhaps that is why, during all those years when the schools were teaching Latin and Greek and totally neglecting science, England produced so few great classical scholars and so many great scientists."[20]

LIBERATION IN A LIBRARY

As the top student of his class at Twyford, Dyson won a scholarship to Winchester College, the boys' high school where his father taught music, and so he moved into the dormitory of the venerated school. Founded in 1382, the school's original buildings were still in use; the place was old and Gothic and grand. Think Hogwarts meets Robin Hood—notched battlements atop fortified stone towers and covered cloisters with fragments of medieval game boards carved into the walls. Indeed, the school's towers and parapets would one day be used in exterior shots in the *Harry Potter* films. For now, in fall 1936, the school marked a new start for Dyson.

In the dorm, boys of all ages were thrown together, with as many as twenty boys sharing a five-hundred-year-old room. Dyson, still a shrimpy *pulex irritans*, cowered in the corner, watching the rowdy older boys. At a place like Winchester College, though, "rowdiness" meant shouting insults in as many obscure languages as possible.[21] Entrance to Winchester was highly competitive, which meant those ancient buildings were full of only the brightest kids. Dyson had found his people. From his cowering corner, he came to idolize an older boy named Frank Thompson, founder of the Winchester College Obscure Languages Club, whose love of poetry inspired Dyson's own.

Unlike Twyford, Winchester College did teach science—but thankfully, just barely. "The teachers were wise enough to leave the kids very much alone," Dyson said. "We had rather few hours in class and we learned more from each other than we did from the teachers, especially in math and science."[22] His chemistry teacher, Eric James, spent class reading poetry aloud: W. H. Auden, Dylan Thomas, Cecil Day Lewis.[23] Meanwhile, Dyson learned chemistry from a classmate—a boy named Christopher Longuet-Higgins.[24] In fact, Christopher, his brother Michael, a boy called James Lighthill, and Dyson became fast friends. The "gang of four," as they referred to themselves, were enamored with science and mathematics and as the newest unofficial chapter of Dyson's secret rebel club, they ventured into that most mysterious place: the library. "It [was] in an old building with the right kind of musty smell of old books and there was there a great treasure house of books which, as far as we knew, nobody was aware of except us," Dyson recalled. "I mean just the four boys."

Among the stacks, Lighthill discovered a heavy, three-volume textbook, *Cours d'analyse de l'école polytechnique*, by one Camille Jordan. It was far too advanced for high school boys—not to mention written entirely in French—all of which contributed to its intrigue for Dyson. How did it get there? Had some other member of the club left it there for them to find?

Together, the boys made their way through all three volumes. "We could not have had a better introduction to serious mathematics," Dyson said. "It went far beyond anything our Winchester teachers knew or cared about."[25] They also dug through all three volumes of Whitehead and Russell's *Principia Mathematica*, a book that spends no fewer than seven hundred pages proving that one plus one equals two. "It was fun to go through," Dyson said. "It gave us a feeling for what the foundations of mathematics were all about."[26] They were fourteen years old.[27]

A book by H. S. M. Coxeter, the great geometer, got them interested in stellating polyhedra. Polyhedra are geometric figures with lots of flat faces that can be "stellated" by extending each face until it meets the plane of another face, making a new polyhedron. Through long winter evenings, the boys painstakingly constructed cardboard models of twenty-sided figures, carefully gluing and painting them until they'd amassed a wondrous collection of origamic shapes and fantastical stars, their own geometric universe.[28] When the boys grew bored of the three-dimensional world, they worked out the vertices of their polyhedra in four.[29] None of this was for class, of course. That was the point—that would have ruined it. "The gang of four learned far more from one another than we learned from our teachers," Dyson said.[30] The key was a potent mixture of friendship and mysterious books—like Ivan Vinogradov's book on number theory, which Dyson, aged sixteen, had to translate into English on his own, learning Russian in the process.[31]

Dyson and Lighthill were each awarded the College's Richardson prize for mathematics—Dyson in 1939 and Lighthill in 1940—but only one

Freeman Dyson examines cardboard models of polyhedra that he and several friends built by hand in the late 1930s, during a visit to Winchester School in May 2009. Photograph by George Dyson.

could win the end-of-term award for the boy with the highest marks on exams.[32] Dyson snagged that one every single time—three times per year, in fact, since each year had three terms. The prize was several shillings, with a stipulation that it be spent on books. Dyson wrote letters to book publishers asking for their catalogs, then ordered books the moment he had his prize money in hand.

He bought *Theoretical Physics* by Georg Joos, Eddington's *New Pathways in Science*, G. H. Hardy's *A Course in Pure Mathematics*, E. T. Whittaker's *Analytical Dynamics*, Felix Klein's *Lectures on the Icosahedron*, David Hilbert's *Grundlagen der Geometrie*, the *Collected Papers* of Srinivasa Ramanujan, plus books by H. G. Wells and a collection of Russian verse.[33] He also bought, with triumphant vindication, E. T. Bell's *Men of Mathematics*—the book his father hadn't let him read—which cemented his desire to pursue mathematics with everything he had. In 1938, for twelve shillings and sixpence-worth of prize money, he ordered *Differential Equations* by H. T. H. Piaggio.[34] He knew it was the kind of math he needed in order to navigate the abstract terrain of Einstein's general relativity, and when the book arrived—small and modestly bound in light-blue cloth—he decided to bring it with him over Christmas vacation.

The Dysons spent their vacations at Oxey Farm, a damp old farmhouse on forty waterlogged acres that they owned on the southern coast of England near the Georgian seaside town of Lymington. George Dyson, who earlier that year left Winchester College to take up the prestigious position of director of the Royal College of Music, felt that his children were too pampered, and so Oxey Farm was intentionally equipped with as few amenities as possible. There was no electricity, and Freeman and Alice had to fill oil lamps to light their way to bed or to read.[35]

But Freeman read. "I started at six in the morning and stopped at ten in the evening, with short breaks for meals," he said. "I averaged

Alice and Freeman Dyson (the latter doing a headstand) at Cemaes Bay in Wales, August 1935. Courtesy of the Dyson family.

fourteen hours a day."[36] He wasn't just reading—the Piaggio book came with seven hundred problems, and Dyson was determined to solve every last one. "I intended to speak the language of Einstein," he said.[37] The vacation lasted a month, and by the time it was ending, he was nearly fluent. But his mother insisted he stop calculating for a little while and join her for a walk.

"*Homo sum: humani nil a me alienum puto*," Mildred intoned to Freeman—"a quotation from the play *The Self-Tormentor* by the African slave Terentius Afer, who became the greatest Latin playwright . . . 'I am human and I let nothing human be alien to me.' This was the creed by which she [Mildred] lived a long and full life until she died at the age of ninety-four," Freeman said. "She told me then, as we walked along the dyke between the mud and the open water, that this should also be my creed. She understood my impatience, and my passion for the abstract beauties of Piaggio. But she begged me not to lose my humanity in my haste to become a mathematician. You will regret it deeply, she said, when one day you are a great scientist and you wake up to find that you have never had time to make friends. What good will it do you to prove the Riemann hypothesis, if you have no wife and no children to share your triumph? You will find even mathematics itself will grow stale and bitter if that is the only thing you are interested in."[38]

The speech had its intended impact on Freeman, who for the rest of his own long life saw the humanity beyond equations and valued people over ideas. Not that it stopped him from finishing all seven hundred of Piaggio's problems.[39] And what Mildred may not have realized that day was that Dyson's love for science and mathematics had never been divorced from friendship. It was the *core* of his friendships and would continue to be the very reason he loved science. "If you are a scientist," he told the Oersted crowd, "you have friends in every corner of the globe. Scientists in every country are linked."[40]

The Dysons didn't know that that vacation would be their last peacetime Christmas, though the signs were there. They were there back in chemistry class when Eric James read from Cecil Day Lewis's 1937 war poem "The Nabara," from a book called *Overtures to Death*.[41] They were there when, in desperation, the Winchester boys' 22-caliber rifles—old, worn-down things they used for target practice—were confiscated for the army.[42] They were there, more and more, in every newspaper, and on every radio, and every time Dyson walked through the war cloister at Winchester College and saw the names of the five hundred former students who'd died in the First World War etched there, he tipped his hat and felt his future light cone shrinking.

"I knew what had happened to the English boys who were fifteen at the start of the First World War and arrived in the trenches in 1917 and 1918," Dyson said. "In all probability I had not many years to live, and every hour spent not doing mathematics was a tragic waste."[43] As the Second World War swung into full effect, many of Winchester's teachers went off to fight, leaving the students to fend largely for themselves.

"I was lucky," Dyson told the Oersted crowd. "When I was halfway through high school the war began and the system began to fall apart. In my last year of high school I spent a total of seven hours a week in class. That was the best time I could have chosen to get an education. . . . Now nobody dreams any more of spending seven hours a week in class. Now the kids are kept chained to their desks and are pumped full of predigested science."[44]

Classes may have fallen apart, but the head of the math department, Clement Durell, saw Dyson's talent, so he brought in a young mathematician, Daniel Pedoe, to be Dyson's personal tutor.[45] Pedoe had been a member of the Institute for Advanced Study in Princeton before returning to England, where he was now teaching at Southhampton University. He gave Dyson problem sets on infinite series—presumably unaware that Dyson had already worked them out in his crib—and was shocked to find that whereas the university students could manage perhaps six problems per homework assignment, Dyson was turning in thirty or forty.

When Dyson requested something more interesting to work on, Pedoe suggested an investigation into the representation of the circles in a plane. Pedoe's own work was in geometry, and he gave Dyson a German translation of a book by Francesco Severi, *Vorlesungen über algebraische*

Geometrie. "I intended it to be weekend browsing," Pedoe recalled. "When we met again . . . it turned out that he had read the whole book and had begun to prove theorems."[46]

"I had to struggle through Severi in German to learn algebraic geometry," Dyson said, "so it was good for my German as well as for my mathematics."[47] Dyson came to consider the book one of his prized possessions.[48] More important, Pedoe was his first encounter with a real live "man of mathematics," like someone out of Bell's book, or at least someone who hung around with such people. It was inspiring. "The lessons with Pedoe were a revelation," Dyson recalled. ". . . I acquired from Pedoe a taste for the geometric style that makes mathematics an art rather than a science."[49]

But as the war days wore on, it became harder to spend one's nights calculating because of the mandatory blackout. Like the rest of England, Winchester was under strict orders that no stray lights emanate from any building, streetlamp, or traffic light in an effort to evade night bombing. In the end, Winchester squeaked by with barely any bombing; rumor had it that Hitler wanted to preserve the ancient city because he fantasized that his "coronation" would be held in Winchester Cathedral, the longest Gothic cathedral in Europe, consecrated in 1093, just a short walk from Dyson's school. Southampton, by contrast, where Pedoe taught, suffered fifty-seven attacks that destroyed more than three thousand buildings and reduced much of the city to a burning rubble whose glow could be seen all the way from France.[50]

While math and science nights were on hold, Dyson found a different activity to occupy him. He and his friend Peter Sankey, another science enthusiast and rebel, took up "night climbing." After midnight, under the cover of enforced darkness with only the moon to light their way, they climbed the ancient buildings of Winchester College, scrambling up over the rooftops and scaling the crumbling stone towers. "We did not bother with such frills as helmets or climbing ropes," Dyson said. "It was wartime."[51] So as the fate of the world hung in the balance, Dyson dangled over a hundred-foot drop clutching a medieval stone saint.[52]

TO CAMBRIDGE

In fall 1941, Cambridge University was eerily empty. Most of the students and younger professors had all gone off to join the war. Dyson, arriving there to attend Trinity College, was, at age seventeen, too young to fight. Prodigy has its upsides too.

The older professors were mostly still around, so Dyson enrolled in lectures by G. H. Hardy (Fourier series), Abram Besicovitch (integration), Paul Dirac (quantum mechanics), and Leopold Pars (dynamics). In Hardy's class, he was one of only two students; the other was James Lighthill, from the Winchester gang of four.[53] In that class, the boys learned something remarkable—and not just Fourier series. Hardy too had attended Winchester, they discovered, and he was the one who had (anonymously) placed *Cours d'analyse de l'école polytechnique*—that three-volume French textbook they'd discovered in the musty stacks, the one that had "opened the gates of mathematics" for them—in the Winchester College library.[54] Turns out that another member of the rebel's club *had* left it there for them to find. Dyson also took a relativity class taught by Arthur Eddington, the light cone coming full circle, where he was one of just three students.[55]

Besicovitch was Dyson's favorite teacher, and Dyson was the teacher's only student. Besicovitch worked at the border between geometry and set theory, and his style of mathematics exerted a strong influence on Dyson's own.[56] Dyson was able to learn so much from him because he rarely did his teaching in the classroom. "I had Besicovitch all to myself," Dyson recalled. "He taught me Russian as well as mathematics. We went regularly for long walks on which only Russian was spoken. He had a billiard table in his living room and we played billiards when the weather was too wet for walking. He gave me problems to work on which were impossibly difficult but taught me how to think. His great work on the geometry of plane sets of points became a model for my own later work in physics."[57]

Still, things were lonely and strange at the deserted university. Even the most reliably magical place, the library, wasn't quite as magical this time around. All of the university's rare books and archives had been shuffled out to the countryside so they'd be protected in case of bombing.[58] On campus, Dyson was appointed "staircase marshal" for his dorm, which meant, as he put it in a letter home to his parents, "I have to look after my staircase, put out bombs and carry out corpses, if a bomb happens to burst within twenty yards of us. All my duties have amounted to so far is trying to get a stirrup-pump [standard equipment for putting out firebombs] mended by plumbers who know nothing about it."[59] He spent many of his nights on fire-watching shifts, alone.

But then he got his hands on a mysterious little book, *The Night Climbers of Cambridge*, written under the pseudonym Whipplesnaith. It was a guidebook—aimed at those who wished to risk limb and law to climb the university buildings at night. Complete with instructions, maps, and photographs, the book cataloged the adventures of the Cambridge night climbers, who, it was quickly becoming clear, formed an invisible secret society, a loose collection of stegophilic rebels, separated by time and space but joined by the shirking of authority and a nocturnal love of heights. In fact, the book was a sequel to an earlier *Roof-Climber's Guide to Trinity*, published in 1900 by the founding member of the club, one Geoffrey Winthrop Young. The *Roof-Climber's Guide* was Young's first book but not his last—he went on to publish several mountaineering books as well as multiple collections of poetry. Whether night climbing inspired poetry or poetry inspired night climbing was hard to say, but Lord Byron was a Cambridge night climber too; he had scaled the Wren Library and was rumored to have made the first ascent of the Great Court Fountain.[60]

Suddenly Dyson realized that when he, along with Peter Sankey, had scaled those ancient stone walls and chapel towers at Winchester College, he had unwittingly taken part in a kind of initiation. Empty as the school's grounds now were, an imperceptible camaraderie awaited him overhead in the dark. Even better, it dawned on him now that Peter Sankey, of all places, was there at Cambridge.

REACHING FOR THE STARS

"No night is too dark for climbing," the guidebook said.

That was good news, since the wartime blackout was still in full effect.[61] Dyson and Sankey consulted the night climbers' guidebook before setting out in the darkness to attempt its climbs. First, they conquered the Gateway Column at the Wren Library. Then they managed the Senate House. Finally, they decided to tackle the New Court clock tower at St. John's.[62]

The New Court clock tower was profoundly ornate but clockless—just four empty faces stared blankly where the clocks should have been. No one knew exactly why. The climb, according to the guidebook, was fifty feet up from the ground to the roof of the building, followed by another forty feet up the tower. Dyson and Sankey studied Whipplesnaith's instructions:

To reach the roof . . . go up the Drain-pipe Chimney, halfway along the outer west wall. Two drain-pipes run from the ground to the battlements in the lee of a buttress set at an angle to the main wall. The right-hand pipe is loose in two places, and should not be trusted too far.[63]

They shimmied up the left-hand pipe, two improbable silhouettes in the moonlight, as the ground dropped away beneath them. It was easy enough not to be too scared—not when a stray bomb could drop out of the sky at any moment anyway. It was a reality Dyson knew too well: he'd stood helpless and watched the student union building, hit by a lost German bomber, burn to the ground.[64]

The pipe runs up past four windows on the left, at the top and bottom of which a stone foot-hold can be found.[65]

They kept climbing. There were bright things ahead. Soon Dyson and Sankey would sit down together and write their own guidebook: *The Night Climbers of Winchester.* It would be their own contribution to the secret rebel club.[66]

The pipe ends three feet short of the battlements. There is no bowl; it just stops dead. The last few inches pass through a ledge, and so one can grasp the top with confidence. A stretch on to the battlement, and one is safely on the roof.[67]

Dyson would go on to have a remarkably bright career. So would the rest of the Winchester gang of four—James Lighthill in fluid dynamics, Christopher Longuet-Higgins in theoretical chemistry, Michael Longuet-Higgins in oceanography. They'd all be elected fellows of the Royal Society.

The two buttresses on either side appear to offer the possibility of chimneying. However, they diverge too much and one slips.[68]

The two small figures stood on the rooftop, the tower looming over them. Dyson's great act of rebellion in that moment was to nurture the seed of an overwhelming optimism—a seed that would blossom into the belief that we live in the most interesting of possible universes, that we can overcome all of it: the limits of our biology, the claustrophobia of our planet, the absoluteness of elsewhere, the heat death of the universe.

One can now reach a ledge to the front and above one's head. A scramble lands one on a sort of terrace, a yard wide and two yards long.[69]

They scrambled.

A clockless circle of stone provides the necessary holds. Thus for the first twelve feet.

Some futures would be darker, though—they could sense it. Like Frank Thompson's. Dyson's poetic idol at Winchester, Thompson would parachute into German-occupied Yugoslavia, attempting to link up with a resistance movement in Bulgaria. He'd be captured there and, according to newspaper reports, held trial, where spectators would gape in amazement when he refused a translator, not knowing that he was the founding member of the Winchester College Obscure Languages Club. He would answer his accusers in flawless Bulgarian: "I am ready to die for freedom." And he did.[70]

The next twelve feet are the most difficult part of the climb. . . . Pillars rear up at the outside corners and are joined to the Tower by an arch of sloping stone, festooned on the upper side with ornamentations that should not be trusted too far.[71]

As for Dyson, he'd soon be sent to High Wycombe for Bomber Command, where the same familiar numbers that had danced around his crib would come to mean other things.

For a climber is a man standing on the edge of an abyss.[72]

Dyson and Sankey perched on the edge. Just two years from now, Sankey would be killed in the battle of Arnhem.[73]

From now on it is merely a wriggle to get on to the arch, and the rest of the way is practically a stone ladder to the pinnacle.[74]

They clung to the crowning spire, triumphant, because what they understood implicitly was what Dyson would someday try to tell the Oersted crowd: that *this* was their education. The books. The secret club. Each other. The Trinity College clock rang out, exultant in the dutiful dark, bells to count the final moments of a childhood. In a minute they'd start the climb back down. For now, they listened to the chimes, and Dyson thought: what a beautiful sound.

NOTES

1. Freeman J. Dyson, "To teach or not to teach," acceptance speech for the 1991 Oersted Medal presented by the American Association of Physics Teachers, 22 January 1991, published in the *American Journal of Physics* 59, no. 6 (June 1991): 491–495, on 491.

2. Dyson, "To teach," 495.

3. Freeman J. Dyson, *Birds and Frogs: Selected Papers, 1990–2014* (Hackensack, NJ: World Scientific, 2015), 31.

4. Paul Spicer, *Sir George Dyson: His Life and Music* (Rochester, NY: Boydell Press, 2014), 129.

5. George Dyson in discussion with the author, 23 December 2020.

6. Spicer, *Sir George Dyson*, 146–148.

7. Thomas Lin, "Freeman Dyson: A 'rebel' without a PhD," *Quanta*, 26 March 2014, https://www.quantamagazine.org/a-math-puzzle-worthy-of-freeman-dyson-20140326.

8. Dyson, *Birds and Frogs*, 31.

9. Kenneth Brower, *The Starship and the Canoe* (Seattle: Mountaineers Books, 1978), quoted in loc. 846 (Kindle).

10. Freeman J. Dyson, "Remarks in response to 'Ad Portas' reception," 20 May 1995, at Winchester College, on 1–2.

11. Freeman J. Dyson, *Disturbing the Universe* (New York: Basic Books, 1979), 191–192.

12. Dyson, "To teach," 492.

13. "Professor Freeman Dyson," *Twyford Life* (magazine of the Twyford Society), no. 16, sec. 3 (July 2020).

14. Freeman J. Dyson, "Reading in Winchester College Library" on Web of Stories, accessed 11 January 2021 at https://www.webofstories.com/play/freeman.dyson/15.

15. Dyson, *Birds and Frogs*, 33.

16. Dyson, "To teach," 492.

17. Dyson, "To teach," 492

18. George Dyson in discussion with the author, 23 December 2020.

19. "Professor Freeman Dyson," *Twyford Life*, 3.

20. Dyson, "To teach," 492.

21. Dyson, *Disturbing the Universe*, 36.

22. Freeman J. Dyson, "Exams and friends at Winchester College" on Web of Stories, accessed 11 January 2021 at https://www.webofstories.com/play/freeman.dyson/9.

23. Freeman J. Dyson, *The Scientist as Rebel* (New York: New York Review of Books, 2006), 13.

24. Freeman J. Dyson, "Inspirational Chemistry Teacher: Eric James" on Web of Stories, accessed 11 January 2021 at https://www.webofstories.com/play/freeman.dyson/13.

25. Dyson, *Birds and Frogs*, 33.

26. Freeman J. Dyson, "Opening the Gates of Mathematics" on Web of Stories, accessed 11 January 2021 at https://www.webofstories.com/play/freeman.dyson/12.

27. T. J. Pedley, "Sir (Michael) James Lighthill. 23 January 1924–17 July 1998," *Biographical Memoirs of Fellows of the Royal Society* 47 (2001): 324–356.

28. Seventy years later, on a return visit to the school in March 2009, Dyson found the whole menagerie on display in a big glass cabinet. The administrators had no idea who made them, or when, but were so taken with the figures that they showcased them anyway. The math instructor who pointed them out to Dyson was delighted to finally solve the mystery.

29. Richard L. Gregory and John N. Murrell, "Hugh Christopher Longuet-Higgins: 11 April 1923–27 March 2004," *Biographical Memoirs of Fellows of the Royal Society* 52 (2006): 149–166.

30. Freeman J. Dyson, Preface to *Selected Papers of Freeman Dyson with Commentary* (Providence, RI: American Mathematical Society, 1996), 3.

31. Freeman J. Dyson, "Falling in Love with Russian" on Web of Stories, accessed 11 January 2021 at https://www.webofstories.com/play/freeman.dyson/11.

32. Gregory and Murrell, "Hugh Christopher Longuet-Higgins," 152.

33. Dyson, *Selected Papers*, 3.

34. Dyson, *Disturbing the Universe*, 11.

35. Spicer, *Sir George Dyson*, 155.

36. Dyson, *Disturbing the Universe*, 13.

37. Dyson, *Disturbing the Universe*, 13.

38. Dyson, *Disturbing the Universe*, 14–15.

39. Freeman J. Dyson, "Piaggio and School Holidays" on Web of Stories, accessed 11 January 2021 at https://www.webofstories.com/play/freeman.dyson/19.

40. Dyson, "To teach," 495.

41. Freeman J. Dyson, *Bombs and Poetry*, Tanner Lectures on Human Values delivered at Brasenose College, Oxford University on 5, 12, 19 May 1982, 93.

42. Freeman J. Dyson, *Infinite in All Directions*, Gifford Lectures given at Aberdeen, Scotland, April–November 1985 (New York: Harper Perennial, 2004), 147.

43. Dyson, *Disturbing the Universe*, 14.

44. Dyson, "To teach," 492.

45. Freeman J. Dyson, "Studying Mathematics at Winchester" on Web of Stories, accessed 11 January 2021 at https://www.webofstories.com/play/freeman.dyson/17.

46. Daniel Pedoe, "In Love with Geometry," *College Mathematics Journal* 29, no. 3 (1998): 170–188, on 179.

47. Dyson, "Studying Mathematics at Winchester."

48. Dyson, *Birds and Frogs*, 34.

49. Dyson, *Birds and Frogs*, 34.

50. "Southampton Blitz 70th anniversary remembered," BBC, accessed 8 January 2021 at http://news.bbc.co.uk/local/hampshire/hi/people_and_places/history/newsid_9241000/9241143.stm.

51. Dwight E. Neuenschwander, ed., *Dear Professor Dyson: Twenty Years of Correspondence between Freeman Dyson and Undergraduate Students on Science, Technology, Society and Life* (Hackensack, NJ: World Scientific, 2016), 201.

52. Neuenschwander, *Dear Professor Dyson*, 201.

53. Jagdish Mehra and Kimball Milton, *Climbing the Mountain: The Scientific Biography of Julian Schwinger* (New York: Oxford University Press, 2000), 239.

54. Freeman J. Dyson, "Opening the Gates of Mathematics" on Web of Stories, accessed 11 January 2021 at https://www.webofstories.com/play/freeman.dyson/12.

55. Freeman J. Dyson, *Maker of Patterns: An Autobiography through Letters* (New York: Norton, 2018), 27.

56. Helen Joyce, "A conversation with Freeman Dyson," *Plus* magazine, 1 September 2003, https://plus.maths.org/content/conversation-freeman-dyson.

57. Dyson, *Birds and Frogs*, 34.

58. "The Second World War," Emmanuel College website, accessed 11 January 2021 at https://www.emma.cam.ac.uk/about/history/ww2/.

59. Dyson, *Maker of Patterns*, 6.

60. Richard Williams, "Cambridge Night Climbing History," talk delivered to the Cambridge Society of Victoria at the Kelvin Club, Melbourne, 21 October 2009.

61. Whipplesnaith, *The Night Climbers of Cambridge* (Cambridge: Oleander Press, 2007 [1937]), 19.

62. Williams, "Cambridge Night Climbing History."

63. Whipplesnaith, *Night Climbers*, 97.

64. Dyson, *Maker of Patterns*, 23.

65. Whipplesnaith, *Night Climbers*, 97.

66. Williams, "Cambridge Night Climbing History."

67. Whipplesnaith, *Night Climbers*, 97.

68. Whipplesnaith, *Night Climbers*, 101.

69. Whipplesnaith, *Night Climbers*, 101.

70. Dyson, *Disturbing the Universe*, 38.

71. Whipplesnaith, *Night Climbers*, 101.

72. Whipplesnaith, *Night Climbers*, 221.

73. Neuenschwander, *Dear Professor Dyson*, 201.

74. Whipplesnaith, *Night Climbers*, 103.

WILLIAM THOMAS

In September 1941, two years into the Second World War, Freeman Dyson moved to Cambridge University as a mathematics student at Trinity College. Seventeen years old, he had a two-year deferment from military service and was painfully aware his peers would soon be risking their lives battling Nazism. Meanwhile, he would cultivate his talents in the hallowed confines where Isaac Newton and other illustrious mathematical minds had worked. In fact, with Cambridge's student population diminished, he came to enjoy unusually close access to eminent older professors who had not been recruited into the war effort. During his studies, he worked and played billiards with mathematicians Abram Besicovitch, G. H. Hardy, and J. E. Littlewood and attended lectures by physicists such as Paul Dirac and Arthur Eddington.[1]

It was while Dyson was at Cambridge that he began a lifelong correspondence with his family. Having been bombed out of their London home, his parents were staying with another family in Reading, just west of the city, though his father often slept in his office at the Royal College of Music while working to keep the institution open.[2] But with the Blitz passed, the war became a more distant concern, exercising its evils on other parts of the world. Germany was focused on Russia, a land campaign was underway in North Africa, U-boats threatened Atlantic supply routes, and Britain sought to use its bomber forces to bring the war to Germany. Cambridge was rarely a target for Germany's own residual bomber raids. Nonetheless, in the early morning hours of 28 July 1942, Dyson had a close call when, across the street from his room, the building housing the Cambridge Union debating society was struck and set ablaze. Although he had been appointed a staircase marshal, responsible for emergency response in his own building, he was forbidden from intervening in the fire. So he and the other staircase marshals stood by watching it burn.[3]

At the beginning of 1943, Dyson had just turned nineteen and was months away from the examinations that would complete his abbreviated education and mark the end of his deferment. On 29 January, he attended a talk by the physical-chemist-turned-novelist C. P. Snow, a Cambridge fellow who was working at the Ministry of Labor, helping to place academic scientists in wartime positions. Dyson told his parents that Snow spoke of a new kind of activity called "operational research." Called OR for short, operational research was distinguished from research done in laboratories inasmuch as it was concerned with using scientific methods to discern how the military could best use its equipment in combat operations. "For this they want the best people," Dyson wrote, "as the job involves not only technical knowledge but the ability to extract information and give advice to generals."[4]

The next day, Dyson met with Snow and other recruiters for about five minutes to discuss what his role would be. Dyson told them he would prefer "something active," and Snow replied that, pending military approval, he would be directed toward one of the best available jobs in OR and that he might need to move away from Britain closer to the fighting. Snow warned Dyson that he was not to be lured into thinking it would be glamorous. Dyson wrote to his parents, "One other thing Dr. Snow said, that all war jobs are absolutely foul, and that no one should go into it with rosy and romantic ideas. So I am not expecting to find my position either congenial or important; but it is difficult to dispel rosy illusions until they are knocked out of you."[5] That summer, Dyson was assigned as a junior staff member in the OR section attached to Royal Air Force Bomber Command and stationed a few miles outside of London.

THE PLACE OF SCIENCE

Dyson would not have been aware that by 1943 OR had gained a decidedly rosy reputation among a group of British scientists who held senior positions in the war effort. Although the phrase "operational research" had not held much significance two years earlier, the work was now regarded as a particularly important bridge linking the developers of military technologies to their users. But more than that, it also brought scientists' rigorous methods to the military's efforts to cut through the fog of war.[6] That Snow was among those who thought OR meant something profound for the place of science in the war is evident from the presentation he gave to the students at Cambridge.

But scientific bridge building is not easy. On a return trip to Cambridge sixteen years later, to deliver the university's prestigious Rede lecture, Snow made this problem into the subject with which his name would become indissolubly linked thereafter. Titling his lecture *The Two Cultures and the Scientific Revolution*, Snow described what he considered to be a deep and damaging divide in British society between scientists and people he referred to as "literary intellectuals." Scientists, he argued, were culturally committed to improvement: they felt "the future in their bones." Nonetheless, they were too often disengaged from literary culture and its humanistic concerns. Snow related his experiences interviewing young scientists during and after the war, recalling, "As one would expect, some of the very best scientists have plenty of energy and interest to spare, and we came across several who had read everything that literary people talk about." He could easily have been referring to Dyson. "But that's very rare," Snow went on, reflecting that when many scientists were probed on books they had read, they might sheepishly confess to having "tried a bit of Dickens." Yet Snow's main concern was the literary culture's ignorance of science, particularly because he understood British politics and its civil service to be dominated by that mentality. He supposed it was Britain's failure to take advantage of the "scientific revolution" that threatened its future.[7]

Snow's views of British culture and polity have been repeatedly criticized. And he himself understood that his lecture presented no clear prescription for how science might function better in society.[8] Certainly he knew it was not a simple matter of putting scientists on top rather than keeping them on tap, as the cliché went. A year after *The Two Cultures*, he elaborated on this issue in a trio of lectures at Harvard University titled Science and Government, in which he told the wartime tale of the then recently deceased science advisers Henry Tizard and Frederick Lindemann. Styling it as a "cautionary story," Snow derided Lindemann, an Oxford University physicist, casting him as out of touch with working scientists, overconfident in his favored gadgets, and quick to leverage his longstanding friendship with Winston Churchill. Tizard, by contrast, was a chemist who spent his career leading and cultivating administrative apparatuses within the British state. In the closed-off world of wartime decision-making, with no appeal to the broader scientific community, Tizard harnessed connections throughout the military and its labs to ensure his advice was grounded in a collective base of knowledge.[9]

To accentuate Tizard's and Lindemann's differences, Snow spotlighted two conflicts between them, both of which bore on Dyson's experience of the war. The first took place in the mid-1930s. The Air Ministry picked Tizard to lead a committee tasked with finding methods of defending Britain against attack by bombers, and he quickly focused on advancing the technology that would later be called radar. However, as a member of the committee, Lindemann pressed hard to advance a broader array of technologies, often behind the committee's back and against others' view of their practicality. Because Churchill was then only a backbench member of Parliament, Lindemann lost that battle and was maneuvered off Tizard's committee. This left Tizard to oversee radar's implementation along the

British coast and its integration into fighter interception tactics in time for it to make a decisive contribution to the Battle of Britain in summer 1940. Tizard's consultative approach to scientific advice was at the height of its influence.[10]

However, after Churchill became prime minister in May 1940, Lindemann quickly rose in prominence as an adviser and was soon given the government post of paymaster general and a peerage, taking the title Lord Cherwell. This shift in power dynamics set the stage for the second conflict. After Britain fended off the Blitz in 1940, it scaled up its bomber attacks on German cities, flying at night to minimize losses to antiaircraft fire and enemy fighters. But because British bombing techniques were not good enough to target German factories in the dark, Lindemann circulated a paper in early 1942 arguing that an intense campaign could devastate the cities as a whole, killing and demoralizing their communities of industrial workers. Tizard argued against this proposition, based not on moral grounds but on rough calculations that British attacks could be nowhere near so devastating as claimed. In his view, any benefits from bombing German cities could not justify the heavy losses bombers still suffered, wasting the lives of their highly trained crews and Britain's scarce industrial production capacity. But Tizard's view could not prevail against the government's and military's entrenched commitment to the bombing policy, and he soon retreated to the presidency of Magdalen College at Oxford. When Dyson came to Bomber Command a year later, there was no longer any serious dispute about the British approach to bombing.[11]

Moving swiftly as he did from the first to the second conflict, Snow built a tragic and morally poignant narrative about the triumph of Lindemann's vices and the suppression of Tizard's virtues. Yet he also elided an important story that took place in between. During this time frame, the influence of individual advisers became less important as, largely through Tizard's guidance, the military services built up an expansive apparatus of scientific analysis and advice around the idea of OR. The idea of operational research traced to the prewar radar effort, when it was understood to be the work scientists did to aid in the new technology's implementation in RAF Fighter Command operations. As these efforts expanded, use of the term evolved to connote any research these scientists did into the military's operational planning, regardless of whether particular technologies were at issue.[12]

Tizard himself grasped the significance of this evolution in July 1941 when he wrote a memorandum to RAF leaders explaining his views on the subject.[13] After that, OR cohered and expanded rapidly within the RAF as existing groups at its fighter and coastal commands were formalized as "OR sections" and a new section was built up at RAF Bomber Command. The establishment of the Bomber Command OR Section came about just as doubts started to arise about the efficacy of British bombing operations.[14] J. D. Bernal, a crystallographer and prominent Marxist intellectual, was studying the effects of bombing for the Ministry of Home Security. He later recalled that he introduced Lindemann to Bomber Command's photographic analysts, who showed him evidence that many bombers never came within miles of their intended targets.[15] Lindemann assigned an assistant to make a more thorough analysis of the photographic evidence, and the resulting report, delivered in mid-August, spurred a reckoning with the situation from Churchill on down.[16] Apprised shortly thereafter of the formation of the Bomber Command OR Section, Lindemann regarded it as a welcome development.[17]

By September 1941, the successful organization of OR in the RAF had already generated considerable enthusiasm among its proponents. In the Science and World Order conference convened by the British Association for the Advancement of Science, Bernal pointed to the wartime experience up to that point, including OR, as indicative that science and society could be more closely coupled in times of peace. He remarked, "With such an integrated body of information, research, development, execution, and control, we have the backbone of a scientific administration, one which is scientific through and through and not merely by the addition of a few eminent scientists."[18] This accorded with the view he had laid out in his 1939 book *The Social Function of Science*, that science could be intelligently planned so that it generated more benefits for society at large.

Meanwhile, the success of OR in the RAF led to its rapid spread through-out the British military. The Admiralty established an OR group by the end of 1941, and the Army and War Office began setting up their own OR organization in 1942. Initially based only in Britain, new groups also began appearing in the Middle East, North Africa, and India, which was doubt-less why Snow had told Dyson his war work might take him overseas.[19]

CALCULATION AND PERCEPTION

As it happened, an overseas assignment was not in Dyson's future. On July 25, 1943, he showed up at Bomber Command headquarters in High Wycombe, Buckinghamshire, just northwest of London. He was billeted with a family in a village about five miles away and bicycled uphill to work

British intelligence officers and their staff from the Royal Air Force and the Women's Auxiliary Air Force at work in the Map Section of the Operations Block at Bomber Command Headquarters near High Wycombe, Buckingham-shire, ca. 1942. Source: Imperial War Museums.

each day, coasting back down at night.[20] Most weeks, he was even able to visit his parents in London, and for this reason, his contemporaneous reflections are not preserved in letters to them as they have been for many other parts of his life.[21] However, in later essays and correspondence, he recalled his time at Bomber Command as an unrelentingly rueful experience.

While the OR Section had been created as part of a conscious effort to bridge the scientific and military components of the war effort, Dyson did not see it that way. It was not for a lack of interest in such lofty matters. In his youth, Dyson had mused on a grand cosmological scale about the nature of humanity and the direction of world affairs. And by chance, just after he arrived at Cambridge, Dyson caught J. D. Bernal fresh off his optimistic talk at the British Association conference when he addressed a mathematical club Dyson had just joined. To this audience, Bernal framed his ideas in terms of the "place of mathematics in a planned society." Reporting back to his parents, Dyson referred to Bernal as "the perpetrator of *The Social Function of Science*," writing, "The trouble with him, as I knew it would be, was that he knows nothing about mathematics, even in its most applied forms. The only good thing about it was that he did proclaim that there is something for mathematicians to do in the statistical side of things; but he did not say anything definite about it."[22]

At Bomber Command, Dyson quickly became buried in the statistical side of Britain's bombing strategy, where he developed a raw appreciation of its procedures, horrors, and failures. He later remembered that the mood was buoyant the day he arrived because the strategy had achieved a rare breakthrough success. The night before, the command had hit the port city of Hamburg with 791 bombers, and for the first time, they had been permitted to deploy "Window," a technique also known as chaff, in which bombers deployed blizzards of aluminum-coated strips to confuse enemy radar. To that point, the RAF had held back on authorizing Window's use for more than a year out of fear that Germany might start using it against Britain. Now it appeared to be a resounding success as only twelve bombers were lost in the raid—an unusually small loss rate for a mission over such a target. Shortly after Dyson arrived, there were two more raids on the city, rounding out what was known as Operation Gomorrah. On 27 July, Bomber Command succeeded for the first time in igniting a firestorm, an oxygen-gorging inferno that multiplied the lethality of the attack, killing in the range of 20,000 people in a single night. It is estimated that perhaps 40,000 people were killed over the course of the Hamburg raids, which was not a fact anyone tried to hide. At the time, a British newsreel boasted that neutral sources were reporting a death toll of 58,000.[23]

Dyson recalled that his first task for the OR Section stemmed from the introduction of Window. He was to draw pictures of the chaff cloud as it developed beneath the bombers, taking into account factors such as the wind, to impress on aircrews how important it was that they stick closely to their formations to avoid being picked out by German radars. However, as he settled into the section, his work turned more toward statistics and applied mathematics. He was assigned to a subsection, ORS2d, which was concerned with discerning the causes of bomber losses, with other parts of ORS2 analyzing related issues such as bomber vulnerabilities and the use of radio countermeasures. ORS1 dealt with bombing efficacy, while other sections dealt with matters such as the use of radar aids to navigation and bombing.[24]

British Avro Lancaster bombers flying in formation over Dresden, Germany, during the Second World War. Source: Alamy Images.

When Dyson arrived at Bomber Command, he was accompanied by two other young scientists with whom he would remain lifelong friends, John Carthy and Mike O'Loughlin.[25] Their arrival contributed to the growth of the command's OR staff to its wartime peak of about fifty members, mainly men but some women.[26] They were given space in crowded, temporary buildings that Dyson remembered as nestled among beech trees that blocked out the sunlight even in summer months.[27] He recalled that members of the section's staff also worked closely with members of the Women's Auxiliary Air Force, who served as photographic interpreters, calculators, technicians, drivers, and secretaries. "They did most of the real work of the ORS," he wrote. "They also supplied us with tea and sympathy. They made a depressing situation bearable. Their leader was Sergeant Asplen, a tall and strikingly beautiful girl whose authority was never questioned."[28]

The leader of the OR Section was Basil Dickins, who was appointed when the section was created and remained in the job for the war's duration. Dickins had received a doctorate in 1929 from the Imperial College of Science and Technology in London and spent his early career as a researcher at the Royal Aircraft Establishment, the Air Ministry's main production center and laboratory. Later in the 1930s, he was part of the effort to integrate radar into fighter interception tactics, and in 1940 he was assigned as secretary for the MAUD Committee, which had been set up to investigate the prospect of building an atomic bomb. About the same time, he started compiling monthly statistical reports for Bomber Command on encounters with enemy aircraft and losses before the OR Section's establishment a year later. After the war, he continued to rise within the military R&D branches of the British civil service, ultimately attaining responsibility for the development and procurement of atomic weapons and guided missiles.[29]

Dyson developed a low opinion of Dickins, but he had a high one of his immediate supervisor, Reuben Smeed. Smeed had earned a doctorate in aeronautical engineering from Queen Mary College in London and taught mathematics at Imperial College before the war. In 1939, he joined the

Air Ministry's radar effort before becoming one of the early members of the Bomber Command OR Section in 1941. After the war, he moved into the new field of transportation research, becoming well known for his theoretical analyses of traffic patterns and early advocacy of congestion pricing.[30] There was a hint of that in the wartime work. Dyson recalled Smeed explaining that keeping bombers in tight formation was generally important for the same reason it was important to sail merchant vessels in convoys: it was easier for the enemy to pick off craft that had strayed from the pack. However, he also related that aircrews were reluctant to stay closely packed because of the possibility of collision. Dyson therefore should try to determine the relative risks of getting shot down out of formation versus of colliding within formation.[31]

Part of the problem involved determining what the causes of bomber losses were, which was made difficult by the fact that most of the time their crews did not return. In a few cases, though, they did, making their way back to Britain after bailing out over enemy territory. Dyson remembered that Smeed would go to London to interview these fortunate individuals about once a week, that it was the part of the job he enjoyed the most, and that he would occasionally bring Dyson with him. Dyson later wrote, "We were not supposed to ask them questions about how they got back, but they would sometimes tell us amazing stories anyway. We were *supposed* to ask them questions about how they were shot down. But they had very little information to give us about that. Most of them said they never saw a fighter and had no warning of an attack. There was just a sudden burst of cannon fire, and the aircraft fell apart around them."[32]

Bomber crews who witnessed the destruction of other bombers did not have much better information. Dyson remembered that they would report seeing explosions in the dark, followed by two flaming objects falling toward the ground, a spectacular sight often corroborated by other witnesses. Although the crew members believed these were collisions, it was impossible to say so with certainty. "Most of the events probably involved single bombers, hit by antiaircraft shells or by fighter cannon fire, that broke in half as they disintegrated," he wrote. More reliable clues might be gleaned from collisions that occurred on training missions over Britain and from collisions that occurred over enemy territory but did not prevent the bombers from returning. Based on the cases over Britain, it could be established that lethal collisions significantly outnumbered nonlethal ones. There were caveats for using the same ratio to calculate the number of lethal collisions from the number of nonlethal ones on bombing missions. Crews might be more eager to bail out over Britain, but with a shorter distance to travel to get home, they might also have a better chance of landing a damaged plane safely. The assumption was that such adjustments might simply cancel each other out.[33]

To gain a handle on the overall problem, Dyson used his mathematical training to develop a collision model. In principle, one could calculate a collision rate for a bomber flying in formation by multiplying the density of bombers by the average relative velocity of two bombers by their "mutual presentation area." That area was defined by the portion of the geometric plane perpendicular to their relative velocity in which a collision could occur, which, Dyson noted, is equivalent to what particle physicists now call a collision cross section. The assumption inherent in the model was that bombers would not see approaching bombers in time to maneuver out of the way, which he supposed was probably the case for night flying over Germany.[34]

The difficulty with this approach was the uncertainty surrounding the numerical values for all three factors. Collisions probably occurred much more often when bombers were in vertical motion than horizontal, but it was hard to establish how vigorously they were executing corkscrew maneuvers to evade antiaircraft fire. The dominant uncertainty surrounded aircraft density. Some crews reported large deviations from their intended track, while others reported no deviation at all. Dyson estimated there to be a factor of ten uncertainty surrounding this value, making his formula useless for predictions but allowing him to set upper bounds on collision rates. Assuming even very tightly packed bombers, he figured that collisions would still account for only a minority of aircraft losses. But based on empirical analysis of available data, it appeared that losses due to collision were in fact smaller still, perhaps accounting for a few dozen losses over the course of some sixty thousand aircraft sorties.[35]

Another problem Dyson tackled was to determine the efficacy of an aircraft-mounted radar system known as Monica in warning bombers of impending tail attack and whether crews were right in switching it off because it often produced false alarms. The trouble in this case was that Monica was installed more often on Halifax bombers than on Lancasters, and Halifax bombers had generally higher loss rates but were also sent on less dangerous missions. To sort through the heterogeneous data, Dyson developed methods that he later noted were similar to what epidemiologists now refer to as meta-analysis. He found that by tabulating results across the different kinds of missions in which Monica was used, it seemed that Monica-equipped aircraft actually had a slightly higher loss rate, albeit within one standard deviation of the overall average rate: the crews were right to turn it off.[36] In any event, by September 1944, Bomber Command ordered Monica removed from all aircraft because a German fighter had been captured that was equipped with a passive receiver, which testing showed was effective in homing in on Monica signals.[37]

Dyson used a similar method to work through the relationship between crew experience and loss rates. Generally, more experienced crews had lower loss rates, which meant that less experienced crews were given less dangerous missions. This fact also happened to be important for motivating crews to face death on mission after mission. Although a bomber typically had more than a 95 percent chance of surviving any given mission, only about a quarter of bomber crews would survive a full thirty-mission tour. Given that grim reality, it was helpful to believe that if a crew survived their first several missions, their odds would improve. However, Dyson found that as the bomber campaign dragged on, the advantage of experience disappeared, suggesting that something was destroying bombers against which there was no effective defense. He only later learned the Germans had equipped fighters with an upward-pointing cannon called Schräge Musik, allowing them to sneak up on bombers from below and shoot them down without ever being seen.

Racked with guilt over his inability to perceive what was happening and save thousands of lives, he regarded it as his and the OR Section's "greatest failure."[38]

DEPARTURE AND RECKONING

Dyson's misery at Bomber Command stemmed partly from his sense that his work did not make much difference. Although some of that sense was linked to the failure to perceive threats such as Schräge Musik, much of it came from his view that it could be nearly impossible to effect change even when he thought he knew how to make a real improvement. This was largely why he grew to detest Basil Dickins, whom he regarded as overly deferential due to his status as a civil servant and unwilling to challenge the obdurate chief of the command, Air Marshal Arthur Harris. Dyson was particularly incensed that Dickins did not push an idea he and Smeed had to remove turrets and gunners from bombers, based on the conclusion that defensive fire was largely ineffective and that with less weight, the bombers could fly faster and reduce their losses.[39]

The failure to persuade Bomber Command to even experiment with turretless bombers contrasted in Dyson's mind with a crusade his friend Mike O'Loughlin waged to have the escape hatches on Lancaster bombers enlarged. O'Loughlin had convinced himself that small hatches were hampering crew members' efforts to bail out and kept pushing through two years and various patches of institutional inertia to see the change implemented. Dyson later reflected, "When we looked around us at the brutalities and stupidities of the Command, I got depressed and Mike got angry. Anger is creative; depression is useless."[40] Dyson's gloom became almost too much to bear during a visit to Bomber Command's Pathfinder Force in January 1944 to study radar countermeasures. The Pathfinder Force was responsible for flying ahead and marking targets for others to bomb, and its crews flew not one but two tours, which less than one-tenth of them would survive. Many crew members were about Dyson's age. "They had faced flaming death thirty times and would face it thirty times more if they were lucky," he recalled. "I had not, and would not. They knew, and I knew that they knew, that I was one of those college-educated kids who found themselves cushy civilian jobs and kept out of harm's way."[41]

After this visit, Dyson decided to enlist in the RAF and take his turn on the bombers, potentially using his mathematical skills as a navigator. However, on a visit to the London apartment his parents were now staying in, his mother dissuaded him, telling him he would be "absolutely hopeless" in that role. "You would get lost every time," he remembered her saying. "Of course I won't argue against your going and getting yourself killed if you think that is the right thing to do. But it would be a terrible waste of an airplane," she chided, knowing how to chart a path through his resolve. Dyson went back to work and not long thereafter, on another visit to his parents, there was an air raid. Lying in bed, contemplating the "overwhelming irrelevance of this game of tit-for-tat bombing," a bomb struck two doors down at the Institut français, blowing in his bedroom window. He remembered that he and his mother went outside to the street to watch the Institut building burn, much as he had the Cambridge Union building a couple of years earlier.[42]

While Dyson's memories of the OR Section provide us with an important look at what life was like at that time, his jaundiced views of Dickins and how decisions were made in Bomber Command have been challenged. Randall Wakelam, the section's most thorough historian, has stressed that bombing missions were built around an intricate architecture of technologies, tactics, and procedures that evolved constantly as new equipment was introduced, as weather and seasons changed, as the war progressed, and as the enemy adapted. In his view, the section provided analysis crucial to making intelligent decisions in such an environment and, based on records of command decision-making, he finds that Dickins and Harris were prepared to approach these issues in an openminded way. Wakelam specifically objects to Dyson's escape-hatch story, relating that the command was consistently concerned about the issue but that swift action was encumbered by the difficulties of making significant changes to aircraft in service.[43]

Regardless, Dyson's discontent rested not only in his sense of the OR Section's inefficacy but also in his understanding that it propped up Britain's bombing effort, which he regarded as futile and always threatening to collapse beneath its catastrophic losses. Despite evidence that bombing cities would not cause the German war effort to collapse or even significantly slow its industry, the command continued to sacrifice the lives of its crews at a rate that actually climbed as the war reached its final stages. Moreover, the command continued to slaughter German civilians in ever-larger numbers, even though, despite its best efforts, it succeeded only in replicating the devastating firestorm at Hamburg once, when it incinerated Dresden in February 1945.[44]

Dyson later reflected on how he became enrolled in such a brutal endeavor through a phenomenon that Hannah Arendt called the "banality of evil" in her own assessment of Nazism. "Through science and technology, evil is organized bureaucratically so that no individual is responsible for what happens," Dyson wrote. "Neither the boy in the Lancaster aiming his bombs at an ill-defined splodge on his radar screen, nor the operations officer shuffling papers at squadron headquarters, nor I sitting in my little office in the Operational Research Section and calculating probabilities, had any feeling of personal responsibility. None of us ever saw the people we killed. None of us particularly cared." He supposed this apathy toward killing had come to him gradually as his youthful idealism was stripped away. First, he accepted a role in the war, then a role in bombing, then his role in bombing civilians on the rationale it might help win the war. When it became clear to him that bombing would not win the war, he thought he might at least save the lives of some crew members, but then he found he could not. "In the last spring of the war I could no longer find any excuses," he wrote. "I had surrendered one moral principle after another, and in the end it was all for nothing."[45]

Searching for solace, Dyson sometimes found it in how other people had responded to those same conditions, such as in O'Loughlin's escape-hatch work, or another friend, physicist Sebastian Pease's work on bombing problems in support of the Normandy invasion.[46] Dyson was also

"Through science and technology, evil is organized bureaucratically so that no individual is responsible for what happens."

struck by a peculiar episode in which he was sent to work with Air Vice Marshal Richard Harrison, who was based at a country house in Suffolk where he commanded a group that tended to take on side missions such as laying mines off the German coast. Dyson was surprised to find that Harrison was uninterested in discussing tactical problems and instead, finding a receptive scientific mind, showed off a collection of silkworms that he kept on mulberry bushes cultivated in greenhouses on the property. Reflecting years later, Dyson supposed Harrison had understood the folly of Bomber Command's enterprise. He wrote, "He could not hope to challenge or change the overall direction of the campaign. The best he could do was what he did, to drive [his group] with a light hand and shield his crews so far as possible from deeds of superfluous bravery. The cultivation of silkworms helped keep him sane."[47]

Dyson's guiding star in his Bomber Command universe was Reuben Smeed, who quickly found Dyson a position at Imperial College when he left the OR Section in August 1945. Decades later, Dyson wrote to Smeed that he had arrived at the command having been steeped in a "rarified atmosphere of pacifist idealism and pure mathematics." He went on, "Under your guidance, I learned to deal with the real world and made some modest contributions to the task you were engaged in. . . . Through all the stupidity and savagery of the bombing campaign, through all the muddle and mendacity of the Command bureaucracy, you kept our objectives clean and clear. You did not let us lose sight of the fact that, if our efforts could in the end only reduce the losses of aircraft by one hundredth, we should still be saving the lives of five hundred young men. You became for me a father figure to whom I could turn for help with my personal miseries and frustrations. In those two years at Bomber Command, I survived the pangs of unrequited love and the bitterness of working for an outfit I detested." Dyson remembered that in Smeed's office, filled with "evil-smelling tobacco smoke," he would often find a sympathetic hearing and sound counsel. "Many a time it was your good humour and common sense that kept my tippy boat from capsizing," he wrote.[48]

After a year at Imperial College, during which he returned to pure mathematics, Dyson took on a fellowship back at Cambridge before obtaining another fellowship that brought him to Cornell University in the United States. First, though, in August 1947, he took an opportunity to attend a meeting of students in Münster in British-occupied Germany to bring himself face-to-face with the survivors of the Allied bombing campaign. Like many other cities in Germany, Münster had been devastated. Describing the scene in a letter home, he noted the ruins of the inner city were "all overgrown with grass and vegetation and one hardly notices them." His impression was of a country eager to show itself resilient and hopeful for the future.[49] Dyson's time was mostly occupied by talks on cultural topics ranging from history to religion, as well as by a packed social itinerary, when the conversation occasionally turned to the war. He related one such conversation in another letter: "I said very little, but the

Germans soon became warmed to their subject and unburdened their hearts without restraint. I have seldom found the Germans so genuinely and obviously happy; a description by the U-boat sailor of what happens when a petrol tanker is torpedoed was given with the most single-minded enthusiasm. It reminded me vividly of the descriptions we used to read at Bomber Command of successful incendiary attacks, and of the elation we felt when such attacks succeeded." Amazingly, they could now all be friends.[50]

Arriving at Cornell in September 1947, six years after his arrival at Cambridge, Dyson set out to forge a new identity for himself. There in secluded Ithaca, New York, he was surrounded by a vibrant coterie of physicists, many of whom had their own feelings of guilt and responsibility, bestowed on them by their work at Los Alamos on the atomic bomb. Each responded in a different way. Hans Bethe, for instance, became an eminent consultant, adviser, and physicist-statesman. Philip Morrison had been to Hiroshima and was already an activist. Having just lost his wife to tuberculosis, Richard Feynman was restless, hedonistic, and eager to move on from his military work.[51] Their examples coupled to Dyson's own experience as he took up the mantle previously worn by scientists such as Tizard, Lindemann, Bernal, and Snow, building a new place for science in a society that had been profoundly changed by war.

NOTES

1. Freeman Dyson, *Maker of Patterns: An Autobiography through Letters* (New York: Norton, 2018), chap. 1, on Eddington, 27.

2. Dyson, *Maker of Patterns*, xv, 23–24.

3. Dyson, *Maker of Patterns*, 6, 23. On the fire, see Cambridge Union Society, "Bicentenary Booklet" (Cambridge, UK: Labute, 2014), 37, https://issuu.com/the cambridgeunion/docs/150128_bicentenary_booklet_v5-1xx_l.

4. Letter dated 30 January 1943, in Dyson, *Maker of Patterns*, 27–28.

5. Dyson, *Maker of Patterns*, 28–29.

6. On perceptions of OR's significance, see William Thomas, "Bureaucratic reformism and the cults of Sir Henry Tizard and Operational Research," in *Scientific Governance in Britain, 1914–79*, ed. Don Leggett and Charlotte Sleigh (Manchester: Manchester University Press, 2016), 45–62.

7. C. P. Snow, *The Two Cultures* (Cambridge: Cambridge University Press, 1998), 10 ("future"), 11–12 (discussion of interviews).

8. On Snow's reception, see the introduction by Stefan Collini in Snow, *Two Cultures*, as well as Guy Ortolano, *The Two Cultures Controversy: Science, Literature, and Cultural Politics in Postwar Britain* (Cambridge: Cambridge University Press, 2009). For a historian's assessment of Snow's argument, see David Edgerton, "C. P. Snow as anti-historian of British science: Revisiting the technocratic moment, 1959–1964," *History of Science* 43 (2005): 187–208.

9. C. P. Snow, *Science and Government* (Cambridge, MA: Harvard University Press, 1961).

10. Following Snow's lecture, this history was subject to further dispute and adjudication in a variety of publications and the subsequent historiography. Key accounts are found in Earl of Birkenhead, *The Professor and the Prime Minister: The Official Life of Professor F. A. Lindemann, Viscount Cherwell* (Boston: Houghton Mifflin, 1962), and Ronald W. Clark, *Tizard* (Cambridge, MA: MIT Press, 1965).

11. On debates over the role of scientists' analyses in setting British bombing policy, see Paul Crook, "Science and war: Radical scientists and the Tizard-Cherwell area bombing debate in Britain," *War and Society* 12 (1994): 69–101.

12. The following account is summarized from William Thomas, *Rational Action: The Sciences of Policy in Britain and America, 1940–1960* (Cambridge, MA: MIT Press, 2015), chaps. 6–8, with key primary sources cited specifically.

13. Memorandum from Henry Tizard to Chief and Vice Chief of the Air Staff, "Operational research," 17 July 1941, Papers of Sir Henry Tizard, HTT 302, Imperial War Museum, London, UK.

14. Key accounts of the OR Section's origins are Basil Dickins, "Operational Research in Bomber Command" (n.d.), chap, 1, Ronnie Shepard Fonds, box 2, Laurier Military History Archive, Wilfried Laurier University, Waterloo, Ontario, Canada, accessed 30 December 2020 at http://lmharchive.ca/the-ronnie-shephard-fonds/operational-research-in-bomber-command/; and Randall T. Wakelam, *The*

Science of Bombing: Operational Research in RAF Bomber Command (Toronto: University of Toronto Press, 2009), chaps. 3 and 4.

15. Letter from J. D. Bernal to C. P. Snow, 11 April 1961, Papers of J. D. Bernal, J.217, Cambridge University Library, Cambridge, UK.

16. The "Butt Report" is a key reference point in the history of the British bombing campaign. See, for example, discussion in Richard Overy, *The Bombing War: Europe 1939–1945* (London: Allen Lane, 2013).

17. Minutes of the 22nd meeting of the Air Fighting Committee, 28 August 1941, AIR 2/8653, National Archives of the UK: Public Record Office, London, UK.

18. J. D. Bernal, "The function of the scientist in government policy and administration," *Advancement of Science* 2 (1942): 14–17.

19. For chronology, see Thomas, *Rational Action*, 68–71.

20. Freeman Dyson, "A failure of intelligence," *MIT Technology Review* (November 2006), https://www.technologyreview.com/2006/11/01/227625/a-failure-of-intelligence/.

21. Dyson, *Maker of Patterns*, 37.

22. Letter dated 19 October 1941, in Dyson, *Maker of Patterns*, 3.

23. Dyson, "Failure of intelligence"; loss figures corroborated in Wakelam, *Science of Bombing*, 140; for newsreel, see "Hamburg Hammered," 12 August 1943, film ID 1087.21, https://www.britishpathe.com/video/hamburg-hammered/.

24. Dyson, "Failure of intelligence"; Dickins, "Operational Research in Bomber Command," chap. 1.

25. Dyson, "Failure of intelligence."

26. Dickins, "Operational Research in Bomber Command," chap. 1 and appendices 1 and 2.

27. Dickins, "Operational Research in Bomber Command," chap. 1; Freeman Dyson, *Weapons and Hope* (New York: Harper & Row, 1984), 117.

28. Dyson, "Failure of intelligence."

29. Wakelam, *Science of Bombing*, 35–36; Margaret Gowing, *Britain and Atomic Energy, 1939–1945* (London: Macmillan, 1964), 45, 48.

30. J. G. Wardrop, "Reuben Smeed, 1909–1976," *Journal of the Royal Statistical Society, Series A* 140 (1977): 570–571. Dyson remembered that his impression of Dickins was formed as early as his first encounter with him; see Freeman Dyson, *Disturbing the Universe* (New York: Harper & Row, 1979), 26.

31. Dyson, "Failure of intelligence."

32. Dyson, "Failure of intelligence."

33. Dyson, "Failure of intelligence."

34. Dyson, "Failure of intelligence."

35. Dyson, "Failure of intelligence."

36. Dyson, "Failure of intelligence"; Dyson generalized his methodology in a paper titled "Notes on the comparison of loss rates," copy provided to the author by Dyson.

37. Dickins, "Operational Research in Bomber Command," chap. 17.

38. Dyson, "Failure of intelligence."

39. Dyson, "Failure of intelligence"; Dyson, *Disturbing the Universe*, 25–26.

40. Dyson, *Disturbing the Universe*, 26–28.

41. Dyson, *Disturbing the Universe*, 19–23.

42. Dyson, *Disturbing the Universe*, 24–25.

43. Wakelam, *Science of Bombing*, esp. 226–230, and on the escape hatch issue, see 38, 150–152.

44. Dyson, *Disturbing the Universe*, chap. 2; on Dyson's views of Dresden and the final phases of the war, see 28–31 and "Failure of intelligence."

45. Dyson, *Disturbing the Universe*, 30–31. On the banality of evil, see Hannah Arendt, *Eichmann in Jerusalem: A Report on the Banality of Evil* (New York: Viking, 1963).

46. Dyson, *Disturbing the Universe*, 31; Dyson, "Failure of intelligence." Pease would later become a leading figure in British nuclear fusion research.

47. Dyson, *Weapons and Hope*, 7–8.

48. Letter from Dyson to Reuben Smeed, 7 August 1976, copy provided to the author by Dyson. On his move to Imperial College, see Dyson, *Maker of Patterns*, 37–38.

49. Letter dated 8 August 1947, in Dyson, *Maker of Patterns*, 43–46.

50. Letter dated 13 August 1947, in Dyson, *Maker of Patterns*, 46–49.

51. Dyson's early impressions of Bethe, Morrison, and Feynman are recorded in letters to his parents reprinted in Dyson, *Maker of Patterns*, chap. 4.

DAVID KAISER

"Yesterday I came up here by train; a lovely trip over the mountains and up the spectacular Lehigh valley," Freeman Dyson wrote to his parents in September 1947, reporting his arrival at Cornell University. "The weather here is grey, damp and cool, but one can see already what a fine panorama of hills and lakes surrounds us." Following an enjoyable cruise on the *Queen Elizabeth* from England to New York, the young student began to settle into his dormitory room, where he would spend the 1947–1948 academic year as a graduate student in Cornell's physics department.[1]

In many ways, it had been a longer journey, from the realm of pure mathematics to theoretical physics. Dyson's mathematical training had begun early, first at the elite and competitive Winchester College school for boys as a teenager, followed by two years at Cambridge University. As a second-year undergraduate at Cambridge, he had begun to publish short papers on number theory and statistics—several of them in *Eureka*, a journal run by the student mathematics club, others in the *Journal of the London Mathematical Society*. His brief stint at Cambridge, however, was interrupted by the war. Beginning in June 1943, at the age of nineteen, he plied his mathematical and statistical skills with the Operational Research Section of the Royal Air Force Bomber Command, busily computing analyses of bomber losses and the efficiency of various bombing strategies. He continued to pursue his mathematical interests on the side when he could, largely as an effort, he later wrote, "to keep me sane amid the insanities of the bombing campaign."[2]

During his time at the Bomber Command, Dyson also began to read Walter Heitler's textbook, *The Quantum Theory of Radiation* (1936), which opened his eyes to "tantalizing hints of fundamental difficulties and unsolved problems" lurking within theoretical physics. He made up his mind that if he could succeed in proving a particular conjecture within number theory, he would become a mathematician, but if he could not produce the proof, then he would make his way into theoretical physics. Three months of intense work followed before Dyson decided he would become a physicist. On the strength of his previous work in mathematics, he won a Trinity Fellowship to return to Cambridge soon after the war, where he quickly found in Nicholas Kemmer an accomplished theoretical physicist and a patient teacher who could help Dyson make the transition. Dyson worked hard under Kemmer's tutelage, but Kemmer remained overburdened with teaching duties, and no one else at Cambridge appeared promising as a potential adviser. A chance encounter with Sir Geoffrey Taylor, a Cavendish physicist who specialized in fluid mechanics and had worked at the wartime Los Alamos laboratory as part of the enormous Manhattan Project, convinced Dyson that he should pursue his studies at Cornell with Hans Bethe. "At the time I hardly knew that Cornell existed," Dyson later recalled. Yet with the help of a Commonwealth Fellowship, he was on his way to Ithaca less than a year later.[3]

Once Dyson arrived in Ithaca, he quickly settled into a routine. His dorm room struck him as "very comfortable and suitable for working in," he explained in a letter to his parents upon arrival. "It is not designed for social life, in fact the rules are 'No cooking, No alcohol, and No women.' . . . At present I am taking meals at the students' cafeteria, which is simply stacked with the most delicious food, and I shall have no difficulty in growing fat on $2 a day."[4] A month later he expanded on his daily routine in a long letter to his mother: "I am living a highly professionalized existence, without any private life to speak of, and wake up in the mornings thinking about mesons and photons, and there is not much one can say about that."[5]

Soon Dyson began to write home about his new supervisor, Hans Bethe. Bethe often joined the graduate students for meals and worked closely with each of them as they struggled to make progress on their research. The German émigré gave a funny first impression. He was "the most unCambridgeish person" Dyson had ever met. "Bethe himself is an odd figure, very large and clumsy and with an exceptionally muddy old pair of shoes," Dyson explained to his parents. "He gives the impression of being very clever and friendly, but rather a caricature of a professor; however, he was second in command at Los Alamos, so he must be a first-rate organiser as well."[6]

Bethe routinely assigned research problems to his students, handing out dissertation topics and postdoctoral projects from the long lists of open problems he continually drew up.[7] To Dyson he assigned the task of improving one of Bethe's own recent calculations, which had appeared in the *Physical Review* one month before Dyson arrived.[8] The assignment changed Dyson's life and the course of modern physics. Prodded by Bethe and soon inspired by Richard Feynman, over the next two years Dyson honed a framework for performing the most precise scientific calculations in human history.

QED AND THE CHALLENGE OF INFINITIES

Bethe had whipped off his calculation in nearly one sitting during an excited train ride from New York City to upstate New York in June 1947. Just before boarding the train, he had attended an invitation-only workshop with about two dozen physicists arranged by J. Robert Oppenheimer, held at the sleepy Ram's Head Inn on Shelter Island, off the north fork of Long Island. Oppenheimer had gathered the physicists at the workshop to focus on longstanding problems with a theory known as quantum electrodynamics (QED).

Twenty years earlier, as leading quantum physicists first cobbled QED together, they had stumbled on several pathologies. Straightforward calculation of nearly any physical process—such as the probability that two electrons would scatter off each other—yielded infinity rather than finite results. Such infinities, or "divergences," made no sense: two electrons might have a high probability to scatter or a low probability, but not an infinite probability. The problem seemed to stem from the heart of quantum theory itself. Werner Heisenberg's famous uncertainty principle, which he published in 1927, stipulated strict trade-offs between the precision with which pairs of quantities could be specified at a given moment. Best known became the uneasy, seesaw balance between position and momentum: the more accurately someone could determine where an object was, the less accurately she could know where it was going, and vice versa. A corollary concerned the energy involved in a given physical process and the time during which the process unfolded. To Heisenberg and his colleagues in the early 1930s, this suggested that elementary particles could—temporarily—violate sacrosanct laws like the conservation of energy, as long as the infractions were set right sufficiently quickly. Particles could spontaneously pop into existence, "borrowing" energy from the vacuum, as long as they were destroyed or reabsorbed on a timescale set by Heisenberg's relation. These short-lived quantum fluctuations became known as "virtual particles."[9]

Given the uncertainty principle, Heisenberg and his colleagues reasoned that such virtual particles should be ubiquitous, even unavoidable. A lonesome electron traveling through space had some chance to spontaneously emit a virtual photon—a single particle associated with the electromagnetic field—and later reabsorb it; the electron's behavior, in other words, would be affected by interactions with its own electromagnetic field. The electron's effective mass (m_{eff}) and charge (e_{eff}) would each be compound quantities, made up of two parts: the "bare" values (m_0 and e_0), which already appeared in physicists' equations, plus quantum corrections (δm and δe) stemming from the inevitable interactions with virtual particles. Yet every time Heisenberg and his colleagues or students tried to calculate the quantum corrections, they found infinity. In principle, virtual particles could be created with any energy whatsoever—including infinite energy—and all such possibilities had to be included within the calculations. As far as the physicists could tell, virtual particles seemed to contribute infinite amounts to a particle's mass and charge.[10]

The problems had lingered, unresolved, as Europe slipped ever more quickly into fascism. Before long, physicists in many parts of the world became preoccupied with wartime projects. Barely two years after the Second World War had thundered to a close, Oppenheimer gathered the small group of physicists together on Shelter Island to try to spark some new ideas about QED and the impasse of the infinities.[11]

Although most of the physicists at the meeting were theorists, Oppenheimer had invited a few experimentalists as well, including Willis Lamb and Isidor Rabi from Columbia University. During the war, Lamb and Rabi had both been immersed in the massive radar project, during which they had mastered new experimental techniques involving hypersensitive electronics. Immediately after the war, they had directed their new skills, along with surplus wartime equipment, toward exquisitely fine-tuned measurements of the behavior of electrons within simple atoms like hydrogen. Beginning in spring 1947, Lamb, Rabi, and their colleagues measured tiny deviations from the expected values for quantities like the energy levels of an electron in a hydrogen atom or the response of an electron in a magnetic field. The discrepancies were so small—about one part in a million—that prewar measurements had missed them entirely. Upon hearing of the new results at Shelter Island, the young theoretical physicist Julian Schwinger, Harvard's wunderkind, wondered aloud whether such deviations might arise from subtle interactions between electrons and virtual photons. Bethe couldn't wait to find out. Scratch pad on his lap, he began scribbling a calculation during his train ride home.[12]

Bethe had mastered a rough-and-ready approach to calculation during the war years at Los Alamos, when time pressures allowed little room for formal niceties. He tackled his new QED calculation with a similar spirit. He ignored the subtleties of Einstein's special theory of relativity and forced various integrals to converge by only considering virtual particles whose energies fell within a finite range. In other words, he ignored by fiat the prewar infinities that had marred all previous calculations in QED—and, in the process, he estimated a theoretical value for the shift of an electron's energy level that was remarkably close to Lamb's recent measurement. Upon his return to Ithaca, Bethe had copies of his manuscript mimeographed and sent to the participants from the now-dispersed Shelter Island meeting.[13]

Bethe's calculation convinced his colleagues to take QED seriously again, but that only raised the stakes to figure out some self-consistent way to perform calculations without resorting to Bethe's various approximations. Bethe set Dyson the task of repeating the calculation while incorporating relativistic effects, a significantly more cumbersome calculation. Dyson dove right in. "I am stimulated to work hard at it by frequent discussions with Bethe, and particularly by the feeling that I am on trial, and upon my success in this job will largely depend the amount of pull I shall have when new jobs come along," Dyson reported to his parents.[14] That same week, Bethe cautioned a different graduate student that "the calculation in the relativistic case is by no means simple. . . . There are approximately twenty different terms which have to be integrated."[15] Nonetheless, Dyson made quick work of the task, plunging in with the skills that he had already developed under Kemmer's direction in Cambridge. After two weeks and several hundred pages of notes, Dyson found that several divergent integrals—which gave rise to the troublesome infinities—exactly canceled each other; he then used a variation of Bethe's trick, substituting a reasonable (though still finite) upper limit to force the remaining integrals to converge. He found results in good agreement with both Bethe's original calculation and the experimental numbers reported by the Columbia experimentalists. Moreover, he had demonstrated to Bethe that he could calculate: whereas Bethe's

Freeman Dyson working on quantum electrodynamics at Cornell, April 1948. Courtesy George Dyson.

hasty, nonrelativistic calculation had taken up only two pages in the *Physical Review*, Dyson's calculation eventually filled ten densely packed pages in the journal.[16].

"Much to my surprise," Dyson wrote home, Bethe suggested that he write up the calculation and send it to the *Physical Review*. Having finished the calculation by late October, Dyson spent several weeks writing the paper—boiling down his tall stack of notes into the essential steps of the calculation—and submitted it in early December. On that same day, Bethe called him into his office and suggested that Dyson spend a second year in the United States; moreover, he should spend it with J. Robert Oppenheimer, who had just moved to the Institute for Advanced Study in Princeton, New Jersey.[17] Bethe explained to the Commonwealth Fellowship administrators why Dyson should receive an extension of his fellowship to cover the second year. After noting that Dyson "is far superior, even to the best American graduate students we have at Cornell," Bethe continued:

There is a very compelling reason for the continuation of Dyson's study in this country. He has become interested in the new developments of Quantum Electrodynamics which have brought back to life the subject that was believed dead for almost 20 years. . . . All this work is going on in the United States at the present time, the main centers being Professor Oppenheimer's school at the Institute of Advanced Studies [*sic*], Harvard and M.I.T., and this department. If Dyson is enabled to spend a year with Oppenheimer, he will broaden his knowledge of this field to which he has already contributed so much, and will be able to spread his knowledge after his return to England.[18]

The pace of it all left Dyson breathless. "All this shows how fundamentally right was the idea that made me change from mathematics to physics, in spite of many discouragements," he wrote to his parents. "I have done nothing in the last two months that you could call very clever or difficult; nothing one tenth as hard as my [Trinity] fellowship thesis; but because the problems I am now dealing with are public problems and all the theoretical physicists have been racking their brains over them for ten years with such negligible results, even the most modest contributions are at once publicised and applauded." The prospect of spending the year at the Institute was enticing as well: "I am not immune from the common disease of celebrity-hunting, and should like to make the acquaintance of Einstein and Weyl and Von Neumann before I go home, which can only be done at Princeton."[19]

TRAVELS THROUGH SPACE AND TIME

In the meantime, a new influence had entered Dyson's life. Amid updates to his parents about his work for Bethe, Dyson's letters also began to mention Cornell's other leading theoretical physicist, Richard Feynman. "Feynman is a man for whom I am developing a considerable admiration," Dyson wrote home in November 1947. "He is the brightest of the young theoreticians here, and is the first example I have met of that rare species, the native American scientist":

He is always sizzling with new ideas, most of which are more spectacular than helpful, and hardly any of which get very far before some newer inspiration eclipses them. His most valuable contribution to physics is as a sustainer of morale; when he bursts into the room with his latest brain-wave and proceeds to expound it with the most lavish sound effects and waving about of the arms, life at least is not dull.

Later in the same letter, Dyson described the scene at a party at the Bethes' house. Their son Henry, then five years old, "was not impressed" by the visitors, "in fact the only thing he would say was 'I want Dick. You told me Dick was coming,' and finally he had to be sent off to bed, since Dick (alias Feynman) did not materialise," Dyson recounted. "About half an hour later, Feynman burst into the room, just had time to say 'so sorry I'm late. Had a brilliant idea just as I was coming over,' and then dashed upstairs to console Henry. Conversation then ceased while the company listened to the joyful sounds above, sometimes taking the form of a duet and sometimes of a one-man percussion band."[20] Soon it was not only Henry Bethe who delighted in Feynman's company. Dyson began to spend more time talking with Feynman, and they became fast friends. Describing Feynman as "half genius and half buffoon, who keeps all physicists and their children amused with his effervescent vitality," Dyson wrote home about his frequent interactions with Feynman—informal chats that roamed from quantum electrodynamics to Feynman's love-life and beyond.[21]

Feynman had attended the Shelter Island workshop in June 1947, which had prompted him—as it had for several other participants—to think hard again about QED. Over the next few months, as Dyson wrote home about Feynman's "brain-waves," Feynman pieced together some ideas about how to evaluate integrals before they diverged to infinity in a way that would respect the relevant requirements of special relativity (something neither Bethe nor Dyson had yet been able to do).[22] But that was just the start. As Bethe had warned, QED calculations tended to include dozens of such integrals, of increasing complexity. Feynman began doodling simple stick-figure line drawings to try to keep track of all the terms. For example, the simplest process by which two electrons would repel each other, Feynman reasoned, would be if one electron shot out a virtual photon, which would later be absorbed by the other electron. Upon emitting the photon—which carried off some energy and momentum—the first electron would recoil backwards, much like a hunter firing a rifle. A little while later, the photon would collide with the second electron, knocking it off course. Feynman sketched the process shown on the top of the accompanying figure. Each element in Feynman's little narrative had a certain likelihood to occur—the first electron moving from point 2 to point 6, the second electron moving from point 1 to point 5, each electron emitting or absorbing the virtual photon, and so on—which Feynman could quantify in an accompanying integral. But the two electrons could scatter in all kinds of other ways, trading more and more virtual photons in more complicated arrangements, and these processes would contribute to the full calculation as well. So Feynman next sketched the nine distinct ways that two electrons could trade two virtual photons back and forth, as shown on the bottom of the figure—each of which could then be translated into its own lengthy mathematical expression.[23]

Dyson found Feynman's idiosyncratic approach inscrutable. "Feynman is a man whose ideas are as difficult to make contact with as Bethe's are easy," Dyson explained to his parents; he found Feynman's diagrams "completely baffling."[24] He was in good company. When Oppenheimer organized a sequel to the Shelter Island meeting, held in April 1948 at a resort in the Pocono mountains north of Philadelphia, he invited Feynman to present his new methods. It did not go well. For one thing, Feynman's talk came late in the day. He followed a marathon eight-hour presentation by Julian Schwinger—the participants paused in the middle for a brief lunch break—who marched through his own new approach to QED. Whereas Schwinger's talk was a polished affair, Feynman bumbled through his half hour while leading figures like Niels Bohr, Paul Dirac, and Edward Teller peppered him with questions. (At one point, convinced that Feynman was hopelessly confused, Bohr strode up to the blackboard, plucked the chalk from Feynman's hand, and delivered an impromptu sermon on the fundamentals of quantum theory.)

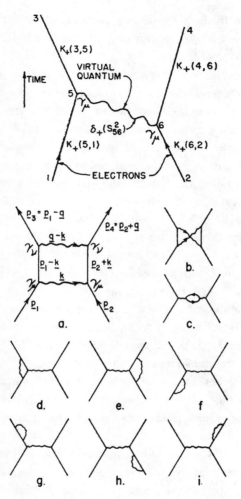

The simplest (top) and next-simplest (bottom) Feynman diagrams for electron scattering. Source: Richard Feynman, "Space-time approach to quantum electrodynamics," *Physical Review* 76 (1949): 772, 787. Courtesy American Physical Society.

Schwinger's methods appeared complicated but rule bound; as far as most listeners at the Pocono meeting could tell, Feynman seemed to make things up as he went along.[25]

Dyson had not been able to attend the Pocono meeting—Oppenheimer refused to invite graduate students, to keep the number of participants from becoming unwieldy—but he had ample opportunities to learn about Feynman's new diagrams that spring. In early June 1948, Dyson received an even more intense dose when he and Feynman drove across the country together, Feynman to visit a girlfriend in Albuquerque and Dyson to take in some sightseeing. Not all of their talk focused narrowly on physics. In the midst of their "Odyssey," as Dyson called it, flooding closed parts of Route 66 in Oklahoma and they had to scramble for lodging. In a breathless, handwritten letter to his parents, Dyson explained that they finally found a room in "what Feynman called a 'dive,' viz. a hotel of the cheapest and most disreputable character, with a notice posted in the corridor saying 'This hotel is under new management, so if you're drunk you've come to the wrong place.'" (In Feynman's less sanitized version, it was a brothel. Feynman made much of how Dyson, "a quiet and dignified English fellow," had been too self-conscious to relieve himself in the sink, since the room had no toilet.) After their road trip, Feynman spent several weeks in New Mexico, where he continued to work on his new diagrammatic calculations, while Dyson made his way by bus to Ann Arbor, Michigan, for the annual summer school in theoretical physics.[26]

The main attraction in Ann Arbor that summer was Julian Schwinger. Following his triumphant day-long presentation at the Pocono meeting, Schwinger strode into town with an entourage of eager acolytes in tow. "Yesterday the great Schwinger arrived, and for the first time I spoke to him," Dyson wrote home in July. "His talks have been from the first minute excellent; there is no doubt that he has taken a lot of trouble to polish up his theory for presentation at this meeting. I think in a few months we shall have forgotten what pre-Schwinger physics was like."[27] The key to Schwinger's approach was to arrange all equations in terms of measurable quantities, such as an electron's effective mass and charge (m_{eff} and

e_{eff}), rather than beginning with "bare" quantities (m_o and e_o) and adding quantum corrections (δm and δe). After all, Schwinger emphasized, physicists could never encounter an electron apart from its cloud of accompanying virtual particles, since there was no way to turn off the uncertainty principle.[28] Divergent quantities like δm and δe would always enter the calculations in combination with the unknown—and unmeasurable— quantities m_o and e_o. If the constants m_o and e_o were also infinite, they could exactly cancel the divergences lurking within δm and δe, leaving unambiguous, finite answers behind. Dyson listened to the lectures in the mornings and worked on "tidying up some of the details of Schwinger's new theory" in the afternoons, which paid some immediate dividends: "I was very glad of this, as it enabled me to speak to Schwinger and the other 'big shots' and get some ideas out of them."[29]

Thus by early August 1948, Dyson—and Dyson alone—had spent intense time working face-to-face with both Feynman and Schwinger, talking in detail about their rival methods. After the summer school ended, Dyson took a bus trip to Berkeley to spend a few weeks relaxing and then began his journey back across the country, again making his way by bus. "On the third day of the journey a remarkable thing happened," Dyson excitedly wrote to his parents: "Going into a sort of semi-stupor as one does after 48 hours of bus-riding, I began to think very hard about physics, and particularly about the rival radiation theories of Schwinger and Feynman":

Gradually my thoughts grew more coherent, and before I knew where I was I had solved the problem that had been in the back of my mind all this year, which was to prove the equivalence of the two theories. Moreover, since each of the two theories is superior in certain features, the proof of the equivalence furnished incidentally a new form of the Schwinger theory which combines the advantages of both.

As soon as he arrived at the Institute for Advanced Study that September, he set about writing up his "magnum opus."[30]

Setting all his thoughts in order proved challenging. "I was for about five days stuck in my rooms, writing and thinking with a concentration which nearly killed me," Dyson wrote home in late September. "On the seventh day the paper was complete, and with immense satisfaction

I wrote the number 52 at the bottom of the last page." He was clearly exhausted: "Having emerged from that, I feel I shall not do any more thinking for the rest of the year." He had managed to demonstrate the mathematical—though by no means conceptual—equivalence between Schwinger's and Feynman's approaches.[31] In fact, as he wrapped up his long paper, Dyson demonstrated a three-way equivalence. Unknown to Schwinger or Feynman before the Pocono meeting, the Japanese theorist Sin-itiro Tomonaga had independently worked out a formal approach to QED in a series of papers and notes stretching back to 1943, remarkably similar to what Schwinger later developed. Unlike Schwinger, Feynman, or their colleagues in the United States, however, Tomonaga had worked out his approach while huddled with his hungry students in a half-burned Quonset hut, their university building having been destroyed by the fire-bombing raids. When Oppenheimer learned of Tomonaga's work right after returning from the Pocono meeting, he immediately mailed copies of several of Tomonaga's papers to the Pocono participants. Bethe shared the package with Dyson, who marveled at the "wonderful demonstration of the resilience of the human spirit."[32]

While writing up his long article, Dyson began to notice several ways to push the new approach beyond what Tomonaga, Schwinger, or Feynman had attempted, so he set to work on a second paper. By Christmas Eve, he had finished nearly sixty pages of the second manuscript, which "turned out to be even more formidable in length and difficulty than the first"; it eventually grew to an eighty-page typescript.[33] Dyson's pair of articles became a how-to guide, enumerating carefully the step-by-step rules for drawing Feynman's new diagrams and explaining how to translate their bare lines into specific mathematical expressions. In this way Dyson, not Feynman, first codified the rules for calculating with the diagrams—precisely what Feynman's frustrated audience had hoped to hear at the Pocono meeting. Even more important, in his second paper, Dyson generalized the previous work. Tomonaga, Schwinger, and Feynman had each developed tricks to isolate and absorb the simplest set of divergences, which sprang up in calculations involving only a single virtual particle. In a stunning tour de force, Dyson demonstrated that the infinities could be removed—systematically—from calculations involving arbitrary numbers of virtual particles. He had transformed QED into a self-consistent theory.[34]

Despite Dyson's careful efforts in his papers to credit Feynman with the invention of the diagrams, many of the earliest diagram users spoke of the "Feynman-Dyson method," the "Feynman-Dyson formulation," and even "Feynman-Dyson diagrams"—and given Dyson's systematic derivation of the diagrams and their use, the label was not entirely inaccurate.[35] At first Dyson found the attribution surprising, even a bit embarrassing, as he described in a letter to his parents after the January 1949 meeting of the American Physical Society in New York:

On the first day the real fun began; I was sitting in the middle of the hall and in the front, with Feynman beside me, and there rose to the platform to speak a young man from Columbia whom I know dimly. The young man had done some calculations using methods of Feynman and me, and he did not confine himself to stating this fact, but referred again and again to "the beautiful theory of Feynman-Dyson," in gushing tones. After he said this the first time, Feynman turned to me and remarked in a loud voice, "Well, Doc, you're in."[36]

The effect on Dyson was immediate: "Since the New York meeting, I have felt some of the consequences of fame," Dyson wrote home excitedly a few weeks later. Oppenheimer had delivered a keynote lecture at the meeting in which he "spoke in enthusiastic terms of the work I have been doing," after which Dyson spent the next twenty-four hours with "one person after another pursuing me and asking to be told all about it." The buzz led to a string of invitations for Dyson to present his work: "beginning in March, a week in Chicago; middle of March, two days at Rochester and a week at Cornell; beginning of April, a week at Toronto; middle of April, the select conference organised by Oppy [a follow-up to the Pocono meeting]; End of April, the annual meeting of the American Physical Society at Washington. With all this, I shall be at Princeton almost as little as Oppy. I have also invitations of a less formal kind from Harvard and Columbia." After his first long paper appeared in print, the avalanche of correspondence from people "asking all kinds of questions, sensible and otherwise," led Dyson to daydream that "soon I shall have to engage a secretary!"[37] His first article appeared in print just days after the January 1949 meeting, months before Feynman published anything on his new methods. Even after Feynman's now-famous papers appeared, physicists continued to cite Dyson's papers more often.[38]

SETTING UP THE "FEYNMAN-SCHOOL"
AT THE INSTITUTE FOR ADVANCED STUDY

Despite the hoopla, Dyson learned quickly that he would need to do more than publish papers in order for the new techniques to spread. Much like his own intense apprenticeship—centered around informal, face-to-face discussions, first with Bethe and then with Feynman and Schwinger—other physicists would need to talk about the diagrams with him, to point, gesture, and doodle, before they could put the new tools to work. Dyson threw himself into that role soon after arriving at the Institute for Advanced Study, quickly transforming from student to mentor.

The Institute had been founded in 1930 near Princeton University but distinct from it. Albert Einstein had become its first full-time member in the early 1930s when he fled Nazi Germany; he was soon joined by dozens of other eminent scholars, many of them similarly displaced by the rise of fascism in Europe. The original plan, articulated by the Institute's inaugural director, had been to provide some refuge where such scholars could "sit and think." ("Well, I can see how you could tell whether they were sitting," one critic quipped.) After the war, a reporter for the *New Yorker* magazine described the "little-vine-covered-cottage atmosphere" at the Institute. Given this history, Hans Bethe had advised Dyson, just before Dyson moved to the Institute in the fall of 1948, "not to expect to find too much going on" there.[39]

On the contrary, Dyson arrived just as Oppenheimer, the Institute's new director, began to change some of the Institute's staid routines. Before accepting the directorship, Oppenheimer had stipulated that he planned to retain some time to pursue physics research—and to surround him-self with an active community of theoretical physicists. During his first year, over objections from long-time members of the Institute, he boosted the number of postdoctoral fellowships for young theoretical physicists there by 60 percent. Oppenheimer's plan, as he explained to a reporter from *Time* magazine in fall 1948, was to convert the Institute into "an intellectual hotel," with a constant hum of visitors passing through.[40] The Institute quickly became a "hotel" that catered to the youth. A *New Yorker* reporter seemed surprised during spring 1949 to find that "most of the Institute's geniuses" seemed "to be in their early twenties and still breathless from taking Ph.D.s on the run."[41]

En route to the Institute—in the midst of his bus trip after the 1948 summer school session in Ann Arbor—Dyson looked forward to his opportunity to impress the famous Oppenheimer.[42] Much to his disap-pointment, Oppenheimer was in Europe when Dyson arrived that Sep-tember and would not return until the middle of October. This turned out to be a blessing in disguise. Dyson joined the largest cohort of young physicists assembled at the Institute to date—the first group to arrive after Oppenheimer's rapid expansion of the postdoc ranks—and the new building that was supposed to hold all the young physicists had not been completed on time.[43] The incoming postdocs therefore all crowded around desks in Oppenheimer's large office while he was away. "We have spent most of our time in argument, physical, political and otherwise," Dyson wrote home a few days before Oppenheimer's return.[44]

With only a single year of graduate study at Cornell under his belt, Dyson at first found it challenging to blend in. He took long walks most afternoons in the Institute's rural surroundings. "Unfortunately my young colleagues are unwilling to join me, as they are obsessed with the Ameri-can idea that you have to work from 9 to 5 even when the work is the-oretical physics. To avoid appearing too superior, I have to say that it is because of bad eyes that I do not work in the afternoons; did you ever hear such nonsense."[45] Before long, though, Dyson struck up some new friendships. One of the new physics postdocs, Jack Steinberger, began to join Dyson on his afternoon walks; another, Cécile Morette, accompanied Dyson on a weekend trip to visit Feynman in Ithaca. With several other postdocs, Dyson enjoyed trips to the movies—he always remarked with awe on the size of his American colleagues' cars—and, after a party with free-flowing whisky, Dyson wrote home that "my relations with these people have been on a much more friendly basis, so the whisky really did me a good service."[46]

As the young physicists got to know each other better, they took turns coaching each other on topics they had focused on during their graduate studies. Dyson—with his rare, firsthand knowledge of Feynman's still-obscure methods—became especially popular with the group.[47] Several of the postdocs had worked with Schwinger at Harvard for their PhDs, and were well versed in his methods; Dyson absorbed all he could from them.[48]

Dyson's enthusiastic lessons about Feynman diagrams to his new friends did not attract Oppenheimer's allegiance upon his return that October. Instead, Oppenheimer expressed frequent opposition to Feynman's and Dyson's work and criticized it bitterly.[49] Feeling "rather like Elijah in the wilderness," Dyson wrote to his parents of his need to "fight" Oppenheimer's "general attitude of hesitation." Oppenheimer allowed Dyson to give a few lectures at the Institute on the new techniques but insisted, in his usual style, on interrupting the talks with cutting criticisms. (At one point, the postdocs asked Dyson to repeat an entire two-hour presentation the following day, without Oppenheimer present.) Unable to compete with Oppenheimer in on-the-spot repartee, Dyson retreated to his typewriter one evening in October to write a memo to his new boss. Feeling "irritable" and stating in the opening line that he "disagree[d] rather strongly with the point of view" that Oppenheimer had been espousing, Dyson reiterated his newfound faith that Feynman's diagrammatic methods were "considerably easier to use, understand, and teach" than either the prewar methods or Schwinger's and Tomonaga's recent approach to QED. Coming to the end of his one-page manifesto, he took a walk to clear his head "and sat down on the grass to make up my mind whether I should send the letter off":

After some time I had finally decided to do it, and then suddenly the sky was filled with the most brilliant northern lights I have ever seen. They lasted only about five minutes, but were a rich blood-red and filled half the sky. Whether the show really was staged for my benefit I doubt, but certainly it produced the same psychological effect as if it had been. I sent the letter off.

The next day Oppenheimer saw Dyson, "said he was delighted with my letter," and offered to give Dyson another opportunity "to expound my views publicly next week." Yet once again, Oppenheimer interrupted Dyson's presentation and adopted a dismissive tone throughout the proceedings. Only after Hans Bethe intervened directly with Oppenheimer a few weeks later did Dyson feel he received some earnest attention. After three more two-hour lectures at the Institute on the new diagrammatic techniques, Oppenheimer ended his criticisms, offering Dyson the simple handwritten note, "Nolo contendere," in November 1948.[50]

From that moment on, all remaining impediments to Dyson's efforts disappeared. By early December, he had convinced several of the postdocs to band together, trying to puzzle through why Feynman, Schwinger, Bethe, and others had gotten different answers when trying to perform the same calculation.[51] Soon groups began undertaking their own diagrammatic calculations, all with Dyson's aid. Wolfgang Pauli captured the interactions perfectly when he wrote to one of the young physicists at the Institute to ask what Dyson and the rest of "the 'Feynman-school'" were working on.[52]

Most significant became Robert Karplus and Norman Kroll's lengthy calculation of the electron's "anomalous" magnetic moment—the tiny experimental anomaly that Isidor Rabi had reported at the 1947 Shelter Island meeting. Schwinger had completed the first self-consistent calculation of the anomaly back in November 1947 but had incorporated only the effects of a single virtual photon. Dyson helped Karplus and Kroll press further, incorporating the effects of two virtual photons in the mix. He handed them a piece of the overall calculation, worked with them until they could rederive that portion, and then coached them through the remaining steps as they evaluated twenty-one distinct Feynman

"What are Dyson and the rest of the 'Feynman-school' working on?"

diagrams, sorted into six classes. Karplus and Kroll announced their result as a demonstration of "the feasibility of Dyson's program." When they added up the contributions from all those complicated diagrams, they found a tiny correction—less than 1.3 percent—to Schwinger's one-photon result.[53] For decades, physicists had struggled to calculate anything in QED and find a finite answer. Now, with Dyson's coaching, young adepts could calculate physical quantities to six decimal places.

With Oppenheimer's blessing and Dyson's active guidance, the Institute fast became a factory of Feynman diagrams. Between 1949 and 1954, the number of articles in the *Physical Review* that used Feynman diagrams grew exponentially, doubling roughly every two years; more than four-fifths of those articles came from members of the Institute's network of diagram users. (The remainder came from Feynman and his own students.) Very quickly, as the postdocs cycled through the Institute, they established a pedagogical cascade: they learned and practiced the diagrammatic techniques during their brief postdoctoral study, then fanned out across the country. Where each of these young physicists went, younger graduate students began taking up the diagrams in their own research; some older physicists began to learn the techniques from their new colleagues as well. In each case, no one at those institutions had used Feynman diagrams in any of their research articles until members of Dyson's Institute network arrived.[54]

Although the cascade spread quickly, it did not reach everywhere at once. Residual pockets remained in which students were unable to learn about the new techniques. As late as October 1952—four and a half years after Feynman's Pocono presentation and more than three years after Dyson's and Feynman's articles on the diagrams had been published—one of Dyson's first converts from the Institute advised an eager graduate student at the University of Pennsylvania that he should either choose a different thesis topic, better fitted to the Penn faculty's strengths, or "get a job and postpone" his thesis altogether. With no members of the Institute's network in town, Dyson's colleague feared that it would be impossible for the student to master the diagrammatic methods.[55]

CONCLUSION

The terms of the Commonwealth Fellowship that had financed Dyson's studies at Cornell and the Institute for Advanced Study stipulated that recipients return to Britain after two years abroad. Dyson obliged, taking up a Royal Society research fellowship at the University of Birmingham in fall 1949; he quickly reestablished ties with his Cambridge adviser, Nicholas Kemmer, as well. Before long, Dyson was training graduate students and postdocs at both universities to use the new diagrammatic methods, much as he had done at the Institute. Just as in the United States, a network of diagram users began to spread throughout Britain.

Several of the new recruits spent time in residence at the Institute for Advanced Study during the early 1950s, reinforcing connections between the groups.[56]

A few months after Dyson moved to Birmingham, Feynman left Cornell for a professorship at Caltech. Faced with Feynman's departure, Bethe hastily arranged for Cornell to hire Dyson in his place. As he wrote to Dyson in spring 1950, "There was a unanimous opinion among the staff" of the Cornell department that "there was only one man in the world who could replace him [Feynman], namely you [Dyson]. Every one of the professors whom I asked for suggestions mentioned your name immediately and spontaneously." Bethe made the case to the Cornell administration, pausing to note simply that "with all his varied activities Mr. Dyson has so far omitted to obtain his doctor's degree"—as if finishing a dissertation were an item one might overlook on a grocery list. The gambit worked: the administration approved the offer, and Dyson accepted,

Freeman Dyson relaxing amid diapers for his firstborn child, Esther, in Zurich, Switzerland, during August 1951, before joining the faculty at Cornell that autumn. Photograph by Verena Huber-Dyson, courtesy George Dyson.

taking up the professorship at the tender age of twenty-six in autumn 1951.[57]

Upon arriving back at Cornell, Dyson began teaching a lecture course on advanced quantum mechanics for the graduate students. His lectures provided the most detailed instructions for performing calculations in QED to date. Three graduate students from the course prepared hektographed copies of Dyson's notes for use by other members of the class. Word of the notes spread, and requests began to come in for more copies. Impressed by the response, the students proceeded to contact physics departments throughout the country to offer copies of the notes for $2.50; Berkeley's physics department, for example, ordered three copies.[58] The notes often served as a proxy textbook for graduate-level courses, as well as a valuable resource for self-study; decades later, a senior physicist at Berkeley pointed to the yellowed notes on his office shelf, calling them simply, "my bible." Dyson's lecture notes were cited dozens of times in research articles and served as a model for the first textbooks on the topic, which began to appear in the mid-1950s and early 1960s.[59]

Soon after starting his professorship at Cornell, Dyson had written home that he found teaching his classes "quite pleasant and I do not work very hard at it." A year later, though, the burden of mentoring large groups of graduate students had taken a toll. "Looking after students for eleven months in the year seems to destroy my freshness and vigor of mind," Dyson lamented in a letter to Bethe. Coaching postdocs informally was one thing; shepherding graduate students through the years-long slog of their dissertations quite another. When Oppenheimer offered Dyson a permanent position at the Institute in 1953, Dyson leaped at the chance.[60]

Although Feynman, Schwinger, and Tomonaga shared the 1965 Nobel Prize in Physics for their work on quantum electrodynamics, it was Dyson who had made QED a workable theory. Generations have since followed Dyson's lead, pushing their diagram-laden calculations to byzantine levels of complexity, all to calculate theoretical predictions about the behavior of particles like electrons to unprecedented accuracy. Back in 1949, Dyson had helped Robert Karplus and Norman Kroll calculate the next-to-leading-order contributions to the electron's magnetic moment by wrangling almost two dozen distinct Feynman diagrams. Young theorists today routinely pursue next-to-next-to-leading-order calculations, and even next-to-next-to-next-to-leading-order calculations—now simply abbreviated as "N^{3}LO"—tackling hundreds and even thousands of diagrams.[61] To date the most audacious calculations continue to involve the electron's magnetic moment. In recent years, teams of physicists have painstakingly computed N^{5}LO corrections, enlisting computer algorithms to help them evaluate contributions from more than fifteen thousand distinct Feynman diagrams. The result of the theoretical calculation agrees with the latest ultrasensitive experimental measurements all the way out to eleven decimal places; the residual mismatch is down to parts per trillion.[62] That's like being able to calculate the distance between the blackboard in Dyson's office at the Institute for Advanced Study and an astronaut's boot print on the moon and getting the answer correct to within the width of a single human hair. Quite a legacy for an impish English transplant with no PhD.

NOTES

Portions of this chapter are excerpted from David Kaiser, *Drawing Theories Apart: The Dispersion of Feynman Diagrams in Postwar Physics* (Chicago: University of Chicago Press, 2005), chap. 3. I am grateful to the University of Chicago Press for permission to reuse this material and to Professor Freeman Dyson for sharing his then-unpublished letters with me during my visit with him in January 2001.

The following abbreviations are used: *FJD*, Freeman J. Dyson letters, originally accessed in Professor Dyson's office at the Institute for Advanced Study in Princeton, New Jersey, now deposited at the American Philosophical Society in Philadelphia, Pennsylvania (filed chronologically, cited by date); *HAB*, Hans A. Bethe papers, collection 14-22-976, Division of Rare and Manuscript Collections, Cornell University Library, Ithaca, New York (cited by box and folder number).

1. Freeman Dyson to his parents, 20 September 1947, in *FJD*; see also Dyson to his parents, 15 September 1947, in *FJD*. Many of Dyson's letters quoted in this chapter were recently published in Freeman Dyson, *Maker of Patterns: An Autobiography through Letters* (New York: Norton, 2018).

2. Freeman Dyson, "Comments on selected papers," in Dyson, *Selected Papers of Freeman Dyson with Commentary* (Providence, RI: American Mathematical Society, 1996), 2–49, on 6; a chronological list of his publications appears in Dyson, *Selected Papers*, 587–595. Dyson described his experiences in the Bomber Command in more detail in Dyson, "Reflections: The sellout," *New Yorker* 46 (21 February 1970): 44–59. On the early development and use of "operations research" by physicists during the Second World War, see Michael Fortun and Silvan Schweber, "Scientists and the legacy of World War II: The case of operations research," *Social Studies of Science* 23 (1993): 595–642; and William Thomas, *Rational Action: The Sciences of Policy in Britain and America, 1940–1960* (Cambridge, MA: MIT Press, 2015). For further biographical details on Dyson's childhood and mathematics education, see Silvan Schweber, *QED and the Men Who Made It: Dyson, Feynman, Schwinger, and Tomonaga* (Princeton: Princeton University Press, 1994), 474–493; and Phillip Schewe, *Maverick Genius: The Pioneering Odyssey of Freeman Dyson* (New York: Thomas Dunne Books, 2013), chap. 1.

3. Dyson, "Comments on selected papers," 8, 10–11. Although he did not prove the Siegel conjecture concerning approximations of numbers by rational fractions, he did manage to strengthen the original statement of the conjecture: Freeman Dyson, "The approximation to algebraic numbers by rationals," *Acta Mathematica* 89 (1947): 225–240, reprinted in Dyson, *Selected Papers*, 65–80.

4. Freeman Dyson to his parents, 20 September 1947, in *FJD*.

5. Freeman Dyson to his mother, 29 October 1947, in *FJD*. Dyson's description of his ascetic lifestyle did not always match other students' impressions of him. Michel Baranger, who arrived at Cornell as a graduate student in September 1949, later recalled the stories that had already attained mythic status around Cornell's physics department. As Baranger heard it, Dyson used to come into the theoretical physics graduate students' office in the mornings, "read the *New York Times*, put his feet on the desk, and then fall asleep and take a nap. In the afternoon he'd go and talk with Bethe, but the other students assumed he wasn't very serious." Naturally, as the story got handed down, the joke was on Dyson's observers as some of his publications began to appear. Michel Baranger interview with the author, 18 January 2001, in Cambridge, Massachusetts.

6. Freeman Dyson to his parents, 25 September 1947, in *FJD*, reprinted in Dyson, *Maker of Patterns*, 57.

7. See Freeman Dyson to his parents, 16 October 1947 and 29 October 1947, in *FJD*. See also Hans Bethe's handwritten notes, "Problems (1/28/48)," in *HAB*, folder 3:42; Hans Bethe, memo to E. Baranger and S. Cohen, 1 February 1952, in

HAB, folder 10:1; and Baranger interview (2001). In a letter of recommendation for one of his students, Bethe explained that "he selected his thesis topic himself, in contrast to practically all our other students." Bethe to Harold Richards, 30 January 1950, in *HAB*, folder 12:12.

8. Hans Bethe, "The electromagnetic shift of energy levels," *Physical Review* 72 (1947): 339–341.

9. On the early history of quantum electrodynamics, see esp. Schweber, *QED*, chaps. 1–2; Arthur I. Miller, "Frame-setting essay," in Miller, *Early Quantum Electrodynamics: A Source Book* (New York: Cambridge University Press, 1994), 1–104; and Kaiser, *Drawing Theories Apart*, 28–37. Many original papers are available in Julian Schwinger, ed., *Selected Papers on Quantum Electrodynamics* (New York: Dover, 1958).

10. Schweber, *QED*, 108–129. Early on, physicists inserted "cutoffs" into their integrals, so that they would only incorporate effects from virtual particles up to some finite energy. The resulting integrals would then remain finite, but the answers would depend on the arbitrary cutoff values.

11. On the 1947 Shelter Island conference, see Schweber, *QED*, chap. 4.

12. Schweber, *QED*, chap. 5; Kaiser, *Drawing Theories Apart*, 37–43.

13. Bethe, "Electromagnetic shift"; Schweber, *QED*, 228–231; Kaiser, *Drawing Theories Apart*, 38–39. On the wartime styles of theorizing, see also Schweber, *QED*, chap. 3; Peter Galison, *Image and Logic: A Material Culture of Microphysics* (Chicago: University of Chicago Press, 1997), chap. 4; and Galison, "Feynman's war: Modeling weapons, modeling nature," *Studies in History and Philosophy of Modern Physics* 29 (1998): 391–434.

14. Freeman Dyson to his parents, 16 October 1947, in *FJD*, reprinted in Dyson, *Maker of Patterns*, 58.

15. Hans Bethe to Marvin Goldberger, 20 October 1947, in *HAB*, folder 10:38; see also Goldberger to Bethe, 13 October 1947, in the same folder.

16. Freeman Dyson to his parents, 16 October 1947 and 29 October 1947, in *FJD*. Dyson mentions the large pile of paper in Dyson, *Disturbing the Universe* (New York: Basic Books, 1979), 54, and discusses Bethe's reaction to his calculation in Dyson, "Comments on selected papers," 12. The results were published in Freeman Dyson, "The electromagnetic shift of energy levels," *Physical Review* 73 (1948): 617–626. Dyson ignored the electron's spin and compared two different cutoff procedures to evaluate formally divergent integrals: inserting the same

value that Bethe had used (equivalent to the rest mass of a single, free electron) versus treating the wave function for a bound-state electron as a packet of free-particle states and using a weighted average of the free-particle energies for the high-frequency cutoff. For more on Bethe's and Dyson's energy-shift calculations, see Schweber, *QED*, 228–231, 497–500.

17. Freeman Dyson to his parents, 29 October 1947 and 7 December 1947, in *FJD*, the latter reprinted in Dyson, *Maker of Patterns*, 63–65. Dyson's article was received at the *Physical Review* on 8 December 1947.

18. Hans Bethe to E. K. Wickman (director, Commonwealth Fund), 28 February 1948, in *HAB*, folder 11:3.

19. Freeman Dyson to his parents, 7 December 1947 and 14 December 1947, in *FJD*.

20. Freeman Dyson to his parents, 19 November 1947, in *FJD*, reprinted in Dyson, *Maker of Patterns*, 59–61. Other colleagues' children also delighted in playing with Feynman: John A. Wheeler, "The young Feynman," *Physics Today* 42 (February 1989), 24–32, on 25; and Tom Kinoshita to the author, 19 March 2003.

21. Freeman Dyson to his parents, 8 March 1948, in *FJD*, reprinted in Dyson, *Maker of Patterns*, 71–72.

22. Richard Feynman, "A relativistic cut-off for classical electrodynamics," *Physical Review* 74 (1948): 939–946; Feynman, "Relativistic cut-off for quantum electrodynamics," *Physical Review* 74 (1948): 1430–1438.

23. See esp. Schweber, *QED*, chap. 8, and Kaiser, *Drawing Theories Apart*, 43–51.

24. Freeman Dyson to his parents, 8 March 1948 and 15 March 1948, in *FJD*, the latter reprinted in Dyson, *Maker of Patterns*, 72–74. Dyson recalls finding Feynman's work "completely baffling" in Dyson, "Comments on selected papers," 12.

25. James Gleick, *Genius: The Life and Science of Richard Feynman* (New York: Pantheon, 1992), 255–261; Schweber, *QED*, 436–445; Kaiser, *Drawing Theories Apart*, 46–47.

26. Freeman Dyson to his parents, 25 June 1948 and 2 July 1948, in *FJD*, reprinted in Dyson, *Maker of Patterns*, 83–90. For more on Dyson's work and friendship with Feynman, see Dyson, *Disturbing the Universe*, 53–68. Feynman's own, and bawdier, recollection of the car trip appears in Richard Feynman with Ralph Leighton, *"What Do You Care What Other People Think?" Further Adventures*

of a Curious Character (New York: Norton, 1988), 65–66. Feynman's original motivation for the road trip had been to visit a girlfriend. He explained in a letter to Bethe when he arrived in Albuquerque, "See, I'm in New Mexico where love has drawn me, but on arrival love dispersed so I am now returned to work." In the meantime, Feynman's work was paying off; he announced to Bethe: "I am the possessor of a swanky new scheme to do each problem in terms of one with one less energy denominator," that is, he had arrived at his trick for combining denominators. Feynman to Hans Bethe, 7 July 1948, in *HAB*, folder 3:42; see also Feynman to Bethe, 22 July 1948, in the same folder, and Dyson to his parents, 8 March 1948, 11 June 1948, and 14 November 1948, in *FJD*.

27. Freeman Dyson to his parents, 22 July 1948, in *FJD*.

28. Hendrik Kramers and Victor Weisskopf had made similar suggestions in the late 1930s, but because they were working with a formalism that did not manifestly respect all the relevant symmetries of special relativity, ambiguities had hampered their calculations. Kramers repeated the suggestion at the 1947 Shelter Island meeting, which resonated with Schwinger. Schwinger had spent the war working on radar, during which he had internalized an effective-circuit approach from his engineering colleagues, in which one focused on input-output variables rather than on the "fundamental" quantities that might appear in the original equations. See Schweber, *QED*, 297–299, 303–340; Galison, *Image and Logic*, 816–827; Kaiser, *Drawing Theories Apart*, 40–42; and Julian Schwinger, "Two shakers of physics: Memorial lecture for Sin-itiro Tomonaga," in *The Birth of Particle Physics*, ed. Laurie Brown and Lillian Hoddeson (New York: Cambridge University Press, 1983), 354–375.

29. Freeman Dyson to his parents, 8 August 1948, in *FJD*. Dyson gave a report on Schwinger's lectures to Bethe, who was in Europe at the time: Dyson to Bethe, 9 August 1948, in *HAB*, folder 3:42.

30. Freeman Dyson to his parents, 14 September 1948, in *FJD*, reprinted in Dyson, *Maker of Patterns*, 98–101. See also Dyson to his parents, 30 September 1948, in *FJD*, reprinted in Dyson, *Maker of Patterns*, 103–105; and Dyson to Bethe, 8 September 1948, in *HAB*, folder 10:20. In his letter to Bethe, Dyson noted that he arrived at his new results "during the enforced idleness of a 60-hour bus-ride from Berkeley to here [Chicago]" and concluded, "Incidentally, the complete equivalence of Schwinger and Feynman is now demonstrated"—the bulk of his letter being occupied instead with how he had succeeded in reformulating Schwinger's methods so as to take advantage of the strengths of Feynman's approach rather than on their equivalence per se. As Dyson saw it, Schwinger's

approach had the advantage of maintaining manifest gauge invariance at each step (as he emphasized in his letter to Bethe of 9 August 1948), whereas Feynman's diagrams provided a highly efficient method for setting up the relevant integrals.

31. Freeman Dyson to his parents, 26 September 1948, in *FJD*. His article appeared as Freeman Dyson, "The radiation theories of Tomonaga, Schwinger, and Feynman," *Physical Review* 75 (1949): 486–502.

32. Freeman Dyson to his parents, 11 April 1948 in *FJD*, reprinted in Dyson, *Maker of Patterns*, 78–81. On Tomonaga's work, see esp. Olivier Darrigol, "Elements of a scientific biography of Tomonaga Sin-itiro," *Historia Scientiarum* 35 (1988): 1–29; Schweber, *QED*, chap. 6; and Kaiser, *Drawing Theories Apart*, 125–143.

33. Freeman Dyson to his parents, 30 September 1948, 21 November 1948, and 24 December 1948, in *FJD*. Dyson's second article appeared as Freeman Dyson, "The S matrix in quantum electrodynamics," *Physical Review* 75 (1949): 1736–1755.

34. Dyson, "S matrix"; see also Schweber, *QED*, 527–549.

35. Kaiser, *Drawing Theories Apart*, 77–78. Dyson sought Bethe's advice while preparing each of his papers, to ensure that proper credit was given to Schwinger and Feynman: Freeman Dyson to his parents, 4 October 1948 and 28 February 1949, in *FJD*, the latter reprinted in Dyson, *Maker of Patterns*, 141–143; Freeman Dyson to Hans Bethe, 10 March 1949, in *HAB*, folder 10:20.

36. Dyson to his parents, 30 January 1949, in *FJD*, reprinted in Dyson, *Maker of Patterns*, 137–139. The speaker was most likely Francis Low, at the time a graduate student at Columbia who was working with Hans Bethe during Bethe's sabbatical year there in 1948: Kaiser, *Drawing Theories Apart*, 79n49.

37. Freeman Dyson to his parents, 30 January 1949, 15 February 1949, 28 February 1949, and 5 April 1949, in *FJD*.

38. On citation patterns, see Kaiser, *Drawing Theories Apart*, 81n58. Dyson's "Radiation theories" paper was received at the *Physical Review* on 6 October 1948 and published in the 1 February 1949 issue; his "S matrix" paper was received on 24 February 1949 and published in the 1 June 1949 issue. Both of Feynman's articles, submitted in April and May 1949, appeared in the 15 September 1949 issue: Richard Feynman, "The theory of positrons," *Physical Review* 76 (1949): 749–759; and Feynman, "Space-time approach to quantum electrodynamics," *Physical Review* 76 (1949): 769–789.

39. On the early years of the Institute for Advanced Study, see Kaiser, *Drawing Theories Apart*, 83–87; Vannevar Bush quoted in "The eternal apprentice," *Time* 52

(8 November 1948): 70–81, on 70 ("Well, I can see"); "Notes and comment," *New Yorker* (30 April 1949), 23–24 ("little-vine-covered-cottage"); Freeman Dyson to his parents, 2 June 1948, in *FJD* ("not to expect").

40. "Eternal apprentice," 81 ("intellectual hotel"). On Oppenheimer's rapid series of reforms at the Institute, see Kaiser, *Drawing Theories Apart*, 86–88.

41. "Notes and comments," 23. Around the same time, a different reporter remarked on "the seeming predominance of youthful people, and the American tempo" at the refurbished Institute: Gertrude Samuels, "Where Einstein studies the cosmos," *New York Times Magazine* (19 November 1950), 14–37, on 14. See also Dyson to his parents, 26 September 1948 and 7 April 1949, in *FJD*.

42. Freeman Dyson to his parents, 14 September 1948, in *FJD*.

43. The group included postdocs Kenneth Case, Cécile Morette, Daniel Feer, Robert Karplus, Norman Kroll, Joseph Lepore, Sheila Power, Fritz Rohrlich, Jack Steinberger, and Kenneth Watson; David Feldman arrived during the second semester. Yukawa Hideki spent that academic year in residence at the Institute, and Abraham Pais, a young theorist who arrived at the Institute in 1947, having spent several months in a Gestapo jail in his native Holland during the war, also worked with the young postdocs. See Kaiser, *Drawing Theories Apart*, 94.

44. Freeman Dyson to his parents, 10 October 1948, in *FJD*.

45. Dyson to his parents, 16 October 1948.

46. Dyson to his parents, 19 October 1948 ("whisky really did me a good service"); see also Dyson to his parents, 10 October 1948, 1 November 1948, and 14 November 1948, in *FJD*.

47. Freeman Dyson to his parents, 4 October 1948 and 10 October 1948, in *FJD*, and Freeman Dyson, interview with the author (8 January 2001), at the Institute for Advanced Study. Norman Kroll later recalled that "I had never understood anything I had heard Feynman say," but upon talking with Dyson soon after both arrived at the Institute that autumn, "I understood what Dyson did very well." Kroll, interview with the author (20 April 2002), in La Jolla, California. Kenneth Case, another of the incoming postdocs, said in a later interview that "the main thing I discovered when arriving at the Institute was there were these peculiar methods that Dick [Feynman] had developed. And nobody had the vaguest idea what they were, except Dyson," who promptly went about explaining it all to the group. Case interview with the author (22 April 2002), in La Jolla, California.

48. Freeman Dyson to his parents, 14 November 1948, in *FJD*.

49. Oppenheimer's criticisms extended beyond the Institute's walls; see in particular his report at the Eighth Solvay Conference, held in Brussels from 27 September to 2 October 1948. In marked contrast with Feynman's "algorithms," Oppenheimer concluded that the methods of Tomonaga and Schwinger had "the advantage of very great generality and a complete conceptual consistency." J. Robert Oppenheimer, "Electron theory," in *Rapports du 8e Conseil de Physique, Solvay*, ed. R. Stoop (Brussels: Solvay, 1950), 1–11, on 7.

50. Freeman Dyson to his parents, 10 October 1948, 16 October 1948, 19 October 1948, 1 November 1948, 4 November 1948, 21 November 1948, and 28 November 1948, in *FJD*. Dyson's memo to Oppenheimer, dated 17 October 1948, is in *FJD* and is reprinted in Dyson, *Maker of Patterns*, 112–113. See also Schweber, *QED*, 520–527, and Dyson, *Disturbing the Universe*, 72–74.

51. Freeman Dyson to his parents, 1 November 1948 and 4 December 1948, in *FJD*.

52. Wolfgang Pauli to Abraham Pais, 26 May 1949, in Pauli, *Wissenschaftlicher Briefwechsel*, ed. Karl von Meyenn (New York: Springer, 1993), 3:655.

53. Julian Schwinger, "On quantum-electrodynamics and the magnetic moment of the electron," *Physical Review* 73 (1947): 416–417; Robert Karplus and Norman Kroll, "Fourth-order corrections in quantum electrodynamics and the magnetic moment of the electron," *Physical Review* 77 (1950): 536–549. See also Freeman Dyson to his parents, 15 February 1949, in *FJD*; Freeman Dyson to Hans Bethe, 10 February 1949, in *HAB*, folder 10:20; and Kaiser, *Drawing Theories Apart*, 209–216. Beginning in the late 1950s, several physicists recomputed the Karplus-Kroll calculation and identified some arithmetic errors in the original analysis: Kaiser, *Drawing Theories Apart*, 215n15. The terms *one-photon* and *two-photon* contributions refer to the number of virtual photons involved in the calculation of corrections to the "bare" electron-photon vertex.

54. Kaiser, *Drawing Theories Apart*, 96–111.

55. Fritz Rohrlich to C. W. Ufford, 13 October 1952, in Fritz Rohrlich papers, Syracuse University Archives, Syracuse, New York, box 23, folder, "Correspondence, 1946–54."

56. Kaiser, *Drawing Theories Apart*, 115–124.

57. Hans Bethe to Freeman Dyson, 5 May 1950 ("unanimous opinion"), and Dyson to Bethe, 23 May 1950, in *HAB*, folder 10:20; Hans Bethe, Lloyd Smith,

and Robert Wilson to Dean L. S. Cottrell Jr., 27 October 1950, in *HAB*, folder 10:20 ("has so far omitted"). Dyson deferred the offer for one year because the Commonwealth Fellowship rules stipulated that recipients could not return to the United States to accept a permanent position for a period of two years. See the correspondence between Dyson and Bethe, ca. 1950–51, in *HAB*, folder 10:20.

58. Hans Bethe to Aage Bohr, 13 March 1953, in *HAB*, folder 10:1. On the preparation and distribution of the notes from Dyson's course, see Stan Cohen, Don Edwards, and Carl Greifinger, form letter dated 9 January 1952, and Rebekah Young to Stan Cohen, 15 January 1952, in Department of Physics records, collection number CU-68, Bancroft Library, University of California, Berkeley, folder 3:31.

59. Kaiser, *Drawing Theories Apart*, 82–83. Dyson's lecture notes have been published: Freeman Dyson, *Advanced Quantum Mechanics*, ed. David Derbes, 2nd ed. (Singapore: World Scientific, 2011 [2007]).

60. Freeman Dyson to his parents, 29 February 1952, in *FJD* ("quite pleasant"); Freeman Dyson to Hans Bethe, 31 August 1952, in *HAB*, folder 10:18 ("Looking after students"). See also Kaiser, *Drawing Theories Apart*, 102n119.

61. See, e.g., K. G. Chetyrkin, B. A. Kniehl, and M. Steinhauser, "Decoupling relations to $O(\alpha_s^3)$ and their connection to low-energy theorems," *Nuclear Physics B* 510 (1998): 61–87, arXiv:hep-ph/9708255; Charalampos Anastasiou, Claude Duhr, Falko Dulat, Elisabetta Furlan, Thomas Gehrmann, Franz Herzog, and Bernhard Mistlberger, "Higgs boson gluon-fusion production at threshold in N3LO QCD," *Physics Letters B* 737 (2014): 325–328, arXiv:1403.4616.

62. Recent theoretical calculations include Tatsumi Aoyama, Masashi Hayakawa, Toichiro Kinoshita, and Makiko Nio, "Tenth-order QED contribution to the electron $g - 2$ and an improved value of the fine structure constant," *Physical Review Letters* 109 (2012): 111807, arXiv:1205.5368; and Aoyama, Hayakawa, Kinoshita, and Nio, "Tenth-order electron anomalous magnetic moment: Contribution of diagrams without closed lepton loops," *Physical Review D* 91 (2015): 033006, arXiv:1412.8284. Compare with the experimental measurement reported in D. Hanneke, S. Fogwell Hoogerheide, and G. Gabrielse, "Cavity control of a single-electron quantum cyclotron: Measuring the electron magnetic moment," *Physical Review A* 83 (2011): 052122, arXiv:1009.4831.

ROBBERT DIJKGRAAF

Freeman Dyson's multifarious life reflected his biggest concern: the importance of diversity. "I look both at scientific and human problems from the point of view of a lover of diversity," he once wrote. "The preservation and fostering of diversity is the great goal which I would like to see embodied in our ethical principles and in our political actions."[1] He famously described himself as a frog, jumping from pool to pool, studying the local details deeply buried in the mud, instead of taking a bird's-eye perspective, soaring high in the sky:

Some mathematicians are birds, others are frogs. Birds fly high in the air and survey broad vistas of mathematics out to the far horizon. They delight in concepts that unify our thinking and bring together diverse problems from different parts of the landscape. Frogs live in the mud below and see only the flowers that grow. They delight in the details of particular objects, and they solve problems one at a time. I happen to be a frog, but many of my best friends are birds.[2]

As a happily hopping frog, Dyson loved to switch academic fields, often after achieving some early signature successes, sometimes jumping to an unfashionable, faraway pool that didn't look very fertile in the eyes of his colleagues. As a lifelong contrarian, it gave him great pleasure to defy expectation, even his own. A saying among his colleagues was that if you wanted Dyson to agree with you, you should surround him with people who disagree with you. As we will see, he employed that amphibian attitude to great effect in his core expertise of mathematical physics, the area where he arguably obtained his most lasting scientific successes, being equally at home in the world of physics as that of mathematics.

This permanent push for intellectual diversity also characterized Dyson's relationship with the institution that would be his academic home for almost seventy years, the Institute for Advanced Study (IAS) in Princeton, New Jersey. It particularly colored his interactions with two high-profile directors who shaped his life as much as he shaped theirs: the theoretical physicist J. Robert Oppenheimer and the economist Carl Kaysen. During his long tenure at IAS, Dyson routinely challenged orthodoxy, encouraged institutional experimentation, and built close personal relationships with his colleagues despite—or sometimes even because of—their intellectual differences.

THE INSTITUTE FOR ADVANCED STUDY

Dyson's first intersection with the Institute for Advanced Study occurred in 1943, in wartime England, when, as a nineteen-year-old student of mathematics at Trinity College, Cambridge, he received an envelope from the United States. As he wrote to his parents, "I was agreeably surprised on Thursday to receive a large envelope stamped Princeton, February 11, 1943, and inside, lo and behold, was *The Consistency of the Continuum Hypothesis*, by Kurt Gödel. This is the first time I have ever been aware, except from an abstract point of view, that a place called America really exists."[3]

Little could he have imagined that five years later, he would be drinking tea with Gödel in his home in Princeton and that within a decade, he would join IAS as a permanent faculty member and become a close colleague of the man who was considered by many as the greatest logician of the twentieth century, if not since Aristotle. Young Dyson did not seem to suffer from much self-doubt, as he went on to comment, "I have been reading the immortal work (it is only sixty pages long) with *The Magic*

Mountain and find it hard to say which is the better." The competition between Kurt Gödel and Thomas Mann, two giants of mathematics and literature, ended in a draw. One could see this balance of science and art as a premonition of Dyson's own career as an outstanding mathematical physicist and beloved author.

Gödel was one of the famous scholars who fled Nazi Germany in the 1930s to find refuge at the IAS, together with such luminaries as theoretical physicists Albert Einstein and Wolfgang Pauli, mathematicians John von Neumann and Herman Weyl, and historians Erwin Panofsky and Felix Gilbert. Probably unknown to Dyson at that time, Thomas Mann had also moved to Princeton, where from 1938 on he was a visiting lecturer at the university. He taught classes on Goethe, Wagner, Freud and, yes, *The Magic Mountain,* and lived at a grand ten-bedroom red-brick house close to the residence of his old friend Einstein.[4]

The Institute, as it is fondly referred to by its inner circles, was the brainchild of its founding director Abraham Flexner, a remarkable educational reformer who had an outsized influence in shaping the course of medicine, scholarship, and philanthropy in the United States in the first half of the twentieth century. The IAS was the physical realization of the platonic vision he had sketched in his famous essay *The Usefulness of Useless Knowledge,* published in 1939 but circulating in draft form since the early 1920s.[5] It was a "paradise for scholars" where scientists and humanists could roam freely and fully concentrate on their research without any teaching obligations or practical applications in mind. Under his stewardship, the Institute soon became a world-leading intellectual center. Working closely together with the Rockefeller Foundation and strongly encouraged by Einstein, it functioned from 1933 on as an intellectual Ellis Island for many European scholars at risk, playing a pivotal role in tilting the intellectual center of gravity across the Atlantic.

The Institute would prove to be an ideal environment for Dyson, allowing him to freely follow his intellectual curiosity and, as a self-described frog, jump from one intellectual pool to another. He holds the current record of the longest-serving faculty member and worked with no fewer than seven directors, as he made me aware of on my very first day on the job as the Institute's director. The admonition of this self-described rebel-scientist was an administrative version of the Hippocratic Oath, "Do no harm." Dyson loved the Institute and had strong ideas of its position in the intellectual landscape.

Dyson first came to Princeton in 1948, during his second year of study in the United States on a Commonwealth Fellowship. He was part of an exceptional group of eight young physicists and mathematicians working with director J. Robert Oppenheimer. In retrospect, that group was remarkably talented and influential. His colleagues included lifelong friends Ken Case, who worked with Oppenheimer in Los Alamos, and Cécile Morette (later DeWitt-Morette, a French mathematician and physicist), the renowned Chinese mathematician Shiing-Shen Chern, and Dyson's first wife, the Swiss logician Verena Huber. This group of young scientists also included two men who would be awarded the Nobel Prize in Physics, the Japanese physicist Hideki Yukawa and the German-American physicist Jack Steinberger.[6] One can argue that Dyson's achievements were only average among this stellar peer group.

Oppenheimer had joined IAS in 1947 as its third director. At that time he was, with the possible exception of his colleague Einstein, arguably the most famous scientist in the country, having successfully led the Los Alamos Laboratory, a key node of the sprawling Manhattan Project, during the war. His move to Princeton captured the headlines of the popular press. He appeared on the cover of *Time* magazine in November 1948 with the somewhat baffling quotation, "What we don't understand, we explain to each other."[7] Soon after that, *Life* magazine devoted a photo essay to the Institute and its new director. He was portrayed as a modern renaissance man, with interests ranging from Sanskrit to sailing in the Caribbean—an image that would stick in the public mind, as the laudation at his honorary degree awarded by Princeton University in 1966 summarized, "physicist and sailor, philosopher and horseman, linguist and cook, lover of fine wine and better poetry."[8]

One of the reasons Oppenheimer considered the directorship of IAS was that he could divide his time as "one-third physics, one-third administration, and one-third policy." The Institute was conveniently situated close to New York and Washington, where he spent much time advising the government on nuclear issues. He not only brought international glamour and prestige to the Institute but also had an ambitious vision of strengthening the rapidly developing field of theoretical particle physics. Returning from Los Alamos, physicists had now set their minds on resolving the mysteries of quantum field theory, a project that was interrupted during the war. Oppenheimer attracted many of these bright minds to Princeton, where they could devote themselves exclusively to research, unburdened by teaching or administrative duties. He put special emphasis on supporting physicists early in their careers, at the postdoctoral stage.[9]

Before the war, the Institute might have had a bit of a staid reputation, a quiet place for established scholars to think deeply where not much happened. Richard Feynman, who was a graduate student at Princeton around that time, observed:

When I was at Princeton in the 1940s I could see what happened to those great minds at the Institute for Advanced Study, who had been specially selected for their tremendous brains and were now given this opportunity to sit in this lovely house by the woods there, with no classes to teach, with no obligations whatsoever. These poor bastards could now sit and think clearly all by themselves, OK? So they don't get any ideas for a while: They have every opportunity to do something, and they are not getting any ideas. I believe that in a situation like this a kind of guilt or depression worms inside of you, and you begin to worry about not getting any ideas. And nothing happens. Still no ideas come.[10]

But now it was suddenly the place to be. Walter Stewart, an economist on the faculty at that time, described the young physicists as "beyond doubt the noisiest, rowdiest, most active and most intellectually alert group we have here."[11]

Oppenheimer quickly appointed several junior physicists to the faculty, all in their early thirties, among them Dyson, Abraham Pais, T. D. Lee, and

C. N. Yang. The latter two would be awarded the Nobel Prize in Physics in 1957, only one year after their theoretical work was published, "for their penetrating investigation of the so-called parity laws which has led to important discoveries regarding the elementary particles."[12] It was a spectacular vindication of Oppenheimer's aggressive policy of chasing excellence in this rapidly moving field.

THE DIVORCE BETWEEN PHYSICS AND MATHEMATICS

Although Dyson's landmark contribution to science was his elegant synthesis of quantum electrodynamics, bringing together the intricate algebra of Julian Schwinger and the intuitive diagrams of Richard Feynman (as described in chapter 3), he was trained as a mathematician in Cambridge, starting his scientific life in the more abstract realms of number theory and combinatorics. He also never obtained a PhD, a fact of which he remained very proud. One can argue that mathematics remained

Freeman Dyson on a drive near Lake Tahoe, California, in 1955. Source: Photograph by Verena Huber-Dyson, courtesy George Dyson.

Dyson's first love throughout his long and rich career, in which he traveled to some of the most remote areas of natural science—and even far beyond that. During all his intellectual wanderings he kept a mathematical perspective, looking to capture complex questions in terms of elegant formulas.

Indeed, one of the most striking aspects of Dyson's scientific life is that he refused to be pigeon-holed as a particle physicist, even though he found early and widespread acclaim in that field, which was arguably the most prestigious branch of science around that time. Many of his colleagues, such as Feynman, Schwinger, and Yang, continued along that road, paved with prizes and public acclaim and leading eventually to the establishment of the successful Standard Model of particle physics.

Characteristically, almost as soon as he had "arrived," Dyson decided to move his formidable analytical skills elsewhere. When his ninetieth birthday was celebrated at the Institute in 2013 with a grand three-day symposium that covered his wide spectrum of interests—from modular forms to climate change, from arms control to extraterrestrial life—the one field that was missing in the program was particle physics.[13]

However, one can argue that it was not so much Dyson leaving particle physics, as particle physics leaving Dyson. From the early 1950s on, after the phenomenal success of QED, the field completely changed its character. It moved from the relative simplicity of atoms to the complexities of nuclear and subnuclear phenomena. New experimental tools allowed scientists to probe nature at smaller distances and higher energies, such as cosmic rays, high-energy particles that come from outer space, and particle accelerators that produce such particles under more controlled laboratory circumstances.

Rapidly, a whole zoo of new particles was discovered—in fact, so many, literally hundreds, that physicists began to suspect that not all of these could be considered elementary. In the struggle to understand this barrage of new subatomic building blocks, one was giving up the hope of finding an underlying logical structure of nature at the smallest scales. Physics seemed to resist any attempts to be captured in concise mathematical formalism. The elegant formalism of quantum electrodynamics—describing all the phenomena of matter and light in terms of just two particles, the electron and the photon—was replaced with a much more descriptive kind of physics.[14] Dyson would call this approach, somewhat dismissively, "phenomenology."

In the mid-1960s, this development came to a boil in the so-called S-matrix or "bootstrap" approach to particle physics. This school considered the collisions in accelerators essentially as a black box. Some particles fly in, others fly out. One can study their correlations, but there is no microscopic description. The black box can never be opened. A bug was made into a feature. The concept of an elementary particle was altogether abandoned. All and none of the particles were elementary. One of the leaders, the Berkeley physicist Geoffrey Chew, wrote that "the aristocratic structure of atomic physics as governed by quantum electrodynamics" should be replaced by "the revolutionary character of nuclear particle democracy. . . . Every nuclear particle should receive equal treatment under the law."[15] One cannot avoid the thought that the social revolution of the 1960s was reflected in particle physics.

All this made the fields of physics and mathematics grow steadily apart. There always has been a certain disdain among physicists about an overly mathematical attitude. Physical intuition, a gut feeling about how nature works, and mathematical rigor, the emphasis on strict proofs according to the rules of logic, are often seen as mutually exclusive. The extra baggage of advanced mathematics only weighs the working physicist down. Feynman famously stated, "If all mathematics disappeared today, physics would be set back exactly one week."[16]

As an indication of this growing cold war between longtime allies, one can consider Feynman's answer to an invitation by his former adviser, John Wheeler, to join a conference in Seattle in 1967 that would bring mathematicians and physicists together to learn from each other. In a long letter, Wheeler encouraged Feynman to share his intuition and not be deterred by abstractions: "Some of us may be able to express ourselves in a language meaningful to mathematicians; others may have to ask for

the forbearance of their colleagues for still talking 'pidgin mathematics.'"
Feynman's answer was characteristically blunt and clear: "Dear John, I am
not interested in what today's mathematicians find interesting. Sincerely
yours, Dick."[17]

As physics became messy and muddy, mathematics turned austere and
rigid. After the war, mathematicians started to reconsider the foundations
of their discipline and turned inside, driven by a greater need for rigor
and abstraction. The more intuitive and approachable style of "old-school"
mathematicians like von Neumann and Weyl, who eagerly embraced new
developments in physics like general relativity and quantum mechanics,
was replaced by the much more austere approach of the next generation.
This movement toward axiomatization and generalization was largely
driven by the French Bourbaki school, started by a group of young mathematicians in Paris in the 1930s. They had set themselves the task of
rebuilding the shabby house of mathematics from its very foundations,
under the motto "structures are the weapons of the mathematician."[18]

One of the leading *bourbakistes* and its de facto early leader was the number theorist André Weil, who joined the Institute in 1958. One characteristic anecdote has Weil's daughter, Nicolette, proudly declaring in school
that her father invented the symbol of the empty set. (Based on the letter Ø in the Norwegian alphabet; Weil read many languages, including
Latin, Greek, and Sanskrit.)[19]

This "purification" of mathematics was strikingly successful. Deep connections among number theory, algebra, and geometry were unearthed
by suitably generalizing elementary concepts such as a point or a curve.
And under the leadership of Weil, the IAS became an intellectual center
of this development.

The growing estrangement of physics and mathematics was reflected in
the academic life at the Institute. Flexner had started IAS with the establishment of a School of Mathematics in 1933. One of his reasons was that
mathematicians are cheap, needing only blackboards and paper, and no
expensive laboratories. But a second reason was that he found a consensus
among all the experts he consulted about who the leading scholars in that

field were. Flexner had a broad view of the mathematical sciences. Einstein was at that time also considered a mathematician. The concept of a theoretical physicist became broadly used only after the Second World War, with Oppenheimer as the prime example that captured the public mind.[20]

In the beginning years, physicists and mathematicians lived peacefully together on the Princeton campus. But under the directorship of Oppenheimer, the two fields began to drift apart. Dyson described it as a divorce:

> We had a unified school of mathematics which included natural sciences; it included Einstein; it included Pauli and other physicists. So, the mathematicians and physicists in the first 20 years of the Institute worked together. But at that time when I became a professor it just coincided with the time when Oppenheimer became director and there was a divorce largely occasioned by the fact that Oppenheimer had no use for pure mathematics, and the pure mathematicians had no use for bombs. So, there was a temperamental incompatibility. For one year, after I became a professor, we still had meetings of the whole faculty, but after that they were separate. We had meetings of the physicists and meetings of the mathematicians, no longer speaking to each other. So, that was sort of a local manifestation of the divorce.[21]

Dyson, though, was happy to be living in both worlds, as he recalled in a later interview:

> But I had the advantage, of course, that I was also a mathematician. And the mathematicians knew that I had done some useful stuff in pure mathematics. And so, when I appeared here as a professor, I was actually invited to the faculty meetings in mathematics. And for several years I used to go to the mathematicians' faculty meetings. And they were quite friendly. They considered me as one of them. . . . So, for the first two years or so I functioned as a mathematician. But at some point, I was, I don't remember how this happened, but I was dropped, and it was made clear that they didn't want me at their meetings. So, they regarded me, at first, as being on their side, but then afterwards they found I wasn't.[22]

The complete administrative separation became official at the end of Oppenheimer's long tenure through the establishment of the School of Natural Sciences in 1965. Disciplines represented in the new school included particle physics, astrophysics, plasma physics, and hydrodynamics. High-energy physicists Stephen Adler and Roger F. Dashen were given five-year appointments that same year and appointed to the faculty in 1969, as was Marshall N. Rosenbluth, a leading theorist in plasma physics, in 1967.

Dyson lamented that separation. In his Josiah Willard Gibbs Lecture, "Missed Opportunities," held under the auspices of the American Mathematical Society in January 1972, he gloomily used the image of a divorce again: "I am acutely aware of the fact that the marriage between mathematics and physics, which was so enormously fruitful in past centuries, has recently ended in divorce."[23] The lecture describes many opportunities in the immediate past where such cross-disciplinary conversations would have proven productive. The Gibbs lecture starts with a quote by the French mathematician Jacques Hadamard that well captures Dyson's own approach to science: "It is important for him who wants to discover not to confine himself to one chapter of science, but to keep in touch with various others."[24]

STABILITY OF MATTER

Having no appetite for the messy phenomenology of particle physics or the abstractions of modern mathematics, Dyson characteristically chose "none of the above" and went a third, independent way. The late 1950s and early 1960s also saw the rise of modern mathematical physics as an independent discipline. It emphasized rigor and advanced mathematical tools such as functional analysis and operator algebra, but it applied these techniques to physical models that had great relevance to the description of the natural world. This field was to a large extent ignored by mainstream mathematicians and physicists, but it provided a wonderfully fertile refuge for Dyson, who made many crucial contributions to the field and added to his prestige.[25]

Oppenheimer was no fan of Dyson's mathematical inclination. When Dyson showed him his rigorous calculation of the ground state energy of a gas of hard spheres, a long-standing and difficult problem in statistical physics, Oppenheimer was not impressed at all and told him bluntly, "You will do better things than this."[26] As Dyson remarked himself, Oppenheimer was right, but not in the way he intended. This work would eventually lead to a series of seminal papers by Dyson and Andrew Lenard that proved rigorously the stability of matter. Indeed, much better work.

In the sciences, mathematical precision has proven particularly relevant in cases where commonsense intuition does not suffice or even leads one astray. Dyson's work on the stability of matter is an excellent example of this kind of work. It answers a simple and general question, "Why doesn't matter collapse?," using sophisticated arguments.

In fact, there is a rich history of the application of mathematics to the general question of stability of natural phenomena. Scientists have long worried about possible catastrophes residing inside the laws of nature. With the birth of the clockwork universe in the seventeenth century, moving according to the laws of Newtonian mechanics, came the great promise that, at least in theory, one could describe with absolute precision the future evolution of everything. Science was seen as a universal machine that takes the present state of the world as input and subsequently computes its future according to precise rules. Science was the ultimate answer to the question, "What happens next?"[27]

But this premise of complete determinism does not prohibit the possibility of hitting a "singularity." The computed answer could be ill defined. This happens to many solutions of mathematical equations. In other words, the clockwork could blow up if asked to predict the future beyond a certain time. For example, following Newton's laws, one of the planets in the solar system could accelerate so fast that its velocity becomes infinite, ejecting it out to infinity. The issue of the stability of the solar system was a particularly vexing problem in the nineteenth century. Apart from the immediate practical concerns, it would signify the breakdown of the very mechanism of science itself.

This classical problem of mechanics was famously settled by the French mathematical giant Henri Poincaré in 1890. By careful analysis of the dynamics, he was able to show that the answer depends crucially on the fine details of the initial state. Generically, the solar system was stable, but for exceptional circumstances, it was not. In a famous episode in the history of mathematics, he found this loophole only after he had submitted a faulty proof of the stability to an international prize competition. To his dismay, Poincaré discovered that tiny variation in the initial condition could lead to a drastically different outcome. This work is now seen as the birth of modern chaos theory, often illustrated by a proverbial butterfly that flips its wings in Brazil and causes a tornado in Texas.[28] (A colleague of mine liked to say that there are also butterflies that prevent a tornado in this fashion, and many lecturers have caused similar havoc by imitating the flapping with their arms.)

With the study of atoms at the beginning of the twentieth century, the question of stability resurfaced, now in the microcosmos. Atoms were first considered as miniature solar systems, with electrons orbiting the central nucleus. This planetary model raised two crucial questions. First, why doesn't an atom self-destroy? Since the electron is a charged particle moving in the electric field of the nucleus, why doesn't it radiate away its energy and spiral inward, crashing into the nucleus? This is called stability of matter of the first kind.

Stability of the second kind concerns the macroscopic behavior of large numbers of atoms. A typical piece of matter contains on the order of a trillion times a trillion atoms. This system can be regarded as a huge collection of negatively charged electrons and positively charged nuclei that interact though attractive and repulsive electric forces. Why do they not all cluster together? Why is the object macroscopic at all? What prevents all the matter in the universe from being able to "shrink into nothing or to diminish indefinitely in size"?[29]

The question of stability of the first kind was answered in the early days of quantum theory by pioneers like Werner Heisenberg and Wolfgang Pauli. Roughly speaking, one shouldn't consider the electrons as miniature planets, but more like waves tightly bound to the nucleus. Just as the waves in a violin string can vibrate only at certain harmonic frequencies, these electron waves are "quantized" and therefore stable. In particular, there is a minimal energy, preventing the formation of a singularity. To further explain why all the electrons in a single atom do not cluster together around the nucleus, one needed a second ingredient: Pauli's exclusion principle. This law says that no two electrons can occupy the same quantum state. When Pauli was awarded the Lorentz Medal in 1931, the physicist Paul Ehrenfest illustrated his principle as follows:

We take a piece of metal. Or a stone. When we think about it, we are astonished that this quantity of matter should occupy so large a volume. Admittedly, the molecules are packed tightly together, and likewise the atoms within each molecule. But why are the atoms themselves so big? . . . Answer: only the Pauli principle, "No two electrons in the same state." That is why atoms are so unnecessarily big, and why metal and stone are so bulky. You will have to concede, Herr Pauli, by partially waiving your exclusion principle you might free us from many worries of daily life, for instance from the traffic problem in our streets.[30]

With the stability of individual atoms well understood, the stability of matter of the second kind was still left open. Dyson was introduced to this problem by the mathematician Andrew Lenard, who visited the Institute from Indiana University in the year 1965–1966. The year before, the mathematical physicists Michael Fisher and David Ruelle had called attention to this important open problem and offered a bottle of champagne for its solution. After a year of hard work, Dyson and Lenard could share the champagne.[31]

What did Dyson and Lenard exactly show? They carefully estimated the overall binding energy due to the electrostatic forces. One can think of this as the energy that is needed to pull all particles apart or, equivalently, as the energy that is released if one assembles matter by bringing the particles together. Matter is stable if there is a maximum value of this binding energy. If such a bound does not exist, the particles can get closer and closer, producing more and more energy. The matter will then collapse in a catastrophic explosion, comparable to the way in which the matter of a star can collapse to form a black hole under the influence of gravity (the outstanding singularity problem in current physics).

To prove stability, one not only needs such an energy bound, but the bound should also scale linearly with the number of particles. Stated otherwise, the maximum binding energy per particle should be fixed and not depend on any external factors. This condition, technically called extensivity, has a very intuitive meaning. Suppose you have two containers, each containing a liter of water. If you add two of these quantities together, you want to end up with two liters of water. If the energy would scale with a higher power of the number of particles, combining two quantities would allow the total content to occupy a smaller volume, releasing excess energy and starting a chain reaction. From this perspective, it is quite a miracle that one can simply pour some more water in a glass, without the water self-destroying, given the complicated rearrangement of charged particles that happens at the microscopic level.

A crucial ingredient in the proof was again the Pauli principle, confirming Ehrenfest's intuition. Dyson and Lenard showed that if you neglect the exclusion principle, the energy of the system would scale very differently, more precisely as the 7/5th power of the number of particles, and therefore would not be stable. In that case, "not only individual atoms but matter in bulk would collapse into a condensed high-density phase. The assembly of any two macroscopic objects would release energy comparable to that of an atomic bomb."[32]

The proof of Dyson and Lenard was convoluted and baroque. Dyson described it as "monstrously long." It is an extended sequence of careful estimates using refined functional analytic methods. The famous mathematical physicist Elliott Lieb, a close colleague of Dyson and professor at Princeton University, described the paper as the most sophisticated application of hard analysis to theoretical physics up to that time. Lieb and his collaborators would go on to drastically improve the work of Dyson and Lennard, with much sharper estimates, shorter proofs, and a clearer conceptual understanding, moving the field to vast generalizations.[33]

RANDOM MATRICES

Another example of the profound impact that Dyson made in mathematical physics was his work on random matrix models. He started working on this subject in the early 1960s, with some key papers published in 1962, but he revisited the subject many times in his long career.

To put this line of research in context, we should distinguish two ways in which mathematical structures appear in the study of nature. The first is in the reduction to fundamental building blocks. Dyson's work on quantum electrodynamics is a good example. It showed how the interaction between light and matter can be systematically approximated by enumerating all elementary processes of electrons and photons. These QED calculations belong to some of the most detailed predictions in science. As noted in chapter 3, the magnetic dipole moment of the electron can be computed, and measured, to a precision of better than one part in a trillion—the equivalent of determining the distance to the moon up to the width of a human hair.

Conversely, in the second way, a fundamental law can "emerge" in the macroscopic limit of great complexity. The laws of thermodynamics, which describe the collective behavior of large collections of molecules, are a perfect illustration of this principle. Only in the limit of an infinite number of molecules do they become mathematical identities.

Random matrices belong in this second category of emerging mathematics. They were introduced by Eugene Wigner, a Hungarian American Nobel Prize laureate in physics and professor at Princeton University, in the description of atomic nuclei. A nucleus is a strongly bound system of protons and neutrons. The dynamics are intricate and impossible to solve exactly. There is no convenient approximation scheme as in the case of QED.

Like any other quantum system, a nucleus is described by a set of states, very much like an atom. Each state has a particular "quantized" energy, and the infinite collection of all such energy levels is called the spectrum. This spectrum can be made visible by exciting the nucleus and observing the emitted radiation. Technically, the spectrum is determined by the Hamiltonian, an infinite matrix that encodes the dynamics of the nucleus. Such a matrix has a set of "eigenvalues," and these eigenvalues correspond to the energy levels of the physical system.

Since it is totally unrealistic to try to compute the exact energy levels of a nucleus, Wigner chose the opposite strategy of complete ignorance. He just picked a random matrix. More precisely, he assumed a Gaussian ensemble of matrices and computed the average properties in this statistical ensemble. (A Gaussian ensemble is one in which properties of individual members are distributed according to the well-known bell-shaped curve.) A key idea in this approach is to first assume that the matrix has a finite rank and subsequently consider the limit where the rank is sent to infinity. Wigner found an elegant result: in this limit, the distance between the individual eigenvalues shrinks to zero, and one obtains a continuous energy spectrum. The eigenvalue density, as a function of the energy, takes the shape of a half-circle, the famous Wigner semicircle law.[34]

Dyson revisited this work in the 1960s and started a much more careful and systematic analysis of these random matrices:

Nuclear physics was a black box. The nucleus was a little black box. You didn't understand what was inside. So, Wigner started thinking of that as a random matrix. And Wigner used that very successfully to describe nuclear reactions, always in a phenomenological sort of way. And I made an axiomatic theory out of that. In 1962 I started thinking about it in a very systematic sort of way. I wrote seven papers which attracted very little attention because they were highly mathematical. But I built this up into a beautiful theory. I enjoyed that more than anything else I've done from the point of view of mathematical beauty. But I had only one disciple, Madan Lal Mehta. . . . He was an Indian who came here. I invited him here for a couple of years. A brilliant mathematician. He had worked on random matrices before, when he was still in France. He emigrated from India to France. He is now a French citizen and lives in Paris. He had written papers about random matrices before I did. I invited him here, and we worked together. It was for me a very good time. That was in '62, '63, '64.[35]

One crucial idea Dyson had was to apply the concept of symmetry. Even in the case of a complex system where most of the details are hidden, there are still general properties that one can assess. For example, does the system preserve the fundamental symmetries of nature?

One such symmetry is time inversion. Flipping the direction of time sounds exotic. It's difficult to unscramble an omelet. But at the level of the elementary particles, most processes have such an invariance. (We ignore here some of the more exotic phenomena in particle physics related to the weak nuclear force.) Dyson's subsequent analysis is elegant. Like so many other things in life, the world of matrix ensembles turns out to be divided into three parts. He baptized this "The Threefold Way."[36] He first distinguished systems with and without time-reversal symmetry. Subsequently he showed, building on a large body of work in mathematics, that in the case that such a symmetry exists, there are two different realizations of the symmetry. His analysis relates the three possibilities directly to the three kinds of number systems (or, more precisely, real division algebras) used in mathematics: real numbers, complex numbers, and quaternions. This elegant and deep paper he considered one of his all-time favorites.[37]

"Frogs live in the mud below and see only the flowers that grow. They delight in the details of particular objects, and they solve problems one at a time. I happen to be a frog."

In the study of random matrices Dyson made good use of his knowledge of condensed matter systems. He showed that the statistical behavior of the energy levels could be reformulated as a collection of two-dimensional electrons interacting through a logarithmic electrostatic force. This also provided a beautiful physical intuition for why the eigenvalues try to avoid each other. The electrons are pushed apart by the mutually repulsive electric force.

Dyson returned to this subject many times during his long scientific career. The theory of random matrices has proven to be remarkably effective in describing many phenomena, from the distribution of parked cars to wireless communications. It is still a growing field, with Dyson's fingerprints all over it.[38] As he recalled in 2004,

I worked ferociously hard for a couple of years and wrote these papers. That fell on stony ground, nobody was interested. Until the last five years or so, and suddenly it's become fashionable again. I was invited to a meeting last year at MSRI [the Mathematical Sciences Research Institute] in Berkeley. . . . It was a whole week of meetings about random matrices, and I was acclaimed as the founding father. And now there's a whole army of young mathematicians who are working on this stuff.[39]

One such surprise application has been to number theory. This was triggered by a legendary discussion that happened over a ritual that has been an important part of scholarly life at the Institute from the very beginning: teatime. Every afternoon, around three o'clock, all the scientists and scholars gather in the common room of Fuld Hall for tea and, more important, freshly baked cookies. The informal discussions that spontaneously arise over that sugar high have triggered some significant breakthroughs in science. For example, it was at a teatime discussion in 1935 that Einstein, in collaboration with Boris Podolsky and Nathan Rosen, came up with the idea of quantum entanglement, the famous "spooky action at a distance" that allows quantum information to be everywhere and nowhere.[40]

In Dyson's case, it was a teatime discussion with the mathematician Hugh Montgomery in April 1972. Montgomery worked on the Riemann hypothesis, widely regarded as the greatest unsolved problem in mathematics. A proof of this conjecture, first formulated by the German mathematician Bernhard Riemann in 1859, unlocks wide areas in number theory, algebra, and geometry. A key ingredient is a mathematical object, the zeta function, that encodes the distribution of prime numbers, the elementary particles of number theory. The Riemann hypothesis makes a strong statement about the zeroes of the zeta function. This is again an infinite collection of real numbers, just like the eigenvalues of a matrix.[41]

Montgomery started to share with Dyson some exciting news: he had studied the statistics of these zeroes and they seemed to avoid each other in a particular fashion. Here Dyson stopped the conversation and immediately suggested a precise formula for that behavior, assuming that the zeroes of the zeta function repel one other in the same way as the energy levels of his beloved random matrices. It turned out he was exactly right. Somehow there is a correspondence between nuclear physics and prime numbers.

This insight has turned out to be a remarkably fertile approach in modern number theory. It did not (yet) lead to a direct proof of the Riemann hypothesis, but this correspondence with random matrices was proven in various generalizations.

In retrospect, Dyson's remarks in the early 1970s about the divorce of mathematics and physics were famous last words, because right at that time, the Standard Model of elementary particle physics was born. In many ways the dream that drove Dyson and others to find the precise formulation of QED using Feynman diagrams was now realized in a vast generalization to include the weak and strong nuclear forces. All the building blocks of that model—gauge fields, curvatures, covariant derivatives—have completely natural mathematical interpretations. One could say that the black box not only can be opened, but inside one finds an elegant mathematical formula. Soon after, mathematicians and physicists started to build a dictionary translating each other's works. I am happy to report that the centuries-old marriage is stronger than ever before. Dyson continued to play an important role in bringing the two partners together, writing many important papers and being an active participant in the weekly mathematical physics brown bag lunch seminar in Princeton.

This does not preclude the two communities to have fundamentally different views of the world. And it is difficult to find a more astute observer of these philosophies than Dyson, who cherished a lifelong belief in the inextricable link between the two disciplines. At the presentation of Graham Farmelo's book *The Universe Speaks in Numbers* at the Institute on 29 May 2019, Dyson remarked, "If physicists and mathematicians visit the zoo, the mathematicians will admire the architecture, the physicists the animals. The big mystery is why the two fit so well."[42]

ROBERT OPPENHEIMER AND DIVERSIFYING THE IAS

Dyson's scientific development cannot be separated from his long and often adversarial relation with J. Robert Oppenheimer, the Institute's director in the defining years 1947 to 1966. That relation was much colored by Dyson's adventurous and contrarian spirit and his permanent quest of intellectual diversity. Oppenheimer was disappointed that Dyson had moved away from particle physics and turned his interests to so many other branches of science. "He thought I was wasting my time, I thought he was a narrow-minded snob, and I told him so."[43]

Oppenheimer didn't hide his feelings either. As Dyson later recalled,

I was doing things that were unworthy of an Institute professor. He had this very exalted idea of particle physics as being the only thing that mattered and was an intellectual snob of course. He thought everything that wasn't pure deep thinking was second rate. I had far more respect for experiments. I was primarily interested in the tools of physics rather than the ideas. He had this very narrow view of what science should be. It's very strange because he had been doing such a great job in Los Alamos where he had to be brought. But he always felt at Los Alamos he was serving his country, but he wasn't really doing science.[44]

Dyson was equally critical about Oppenheimer's direct engagement with current research. He even accused him of being somewhat of a physics poseur:

He hadn't any kind of detailed interest in what people were doing, and he was remote, and the real problem was he was spending about two thirds of the time in Washington and he was so absorbed in public affairs. So for him—I mean he liked to come to the seminars and show off how much he knew about things that he didn't really know. But he never gave us really any sort of leadership. So that was a big disappointment.[45]

In Dyson's view, Oppenheimer, in his Institute years, lacked the patience and staying power for deep dives into physics:

Why did he not succeed in scientific research as brilliantly as he succeeded in soldiering and administration? I believe the main reason why he failed was a lack of *Sitzfleisch*. *Sitzfleisch* is a German word with no equivalent in English. The literal translation is "Sitflesh." It means the ability to sit still and work quietly. He could never sit still long enough to do a difficult calculation. His calculations were always done hastily and often full of mistakes.[46]

Dyson kept pushing Oppenheimer to diversify the group of physicists at IAS beyond particle theory. For example, he argued strongly for the appointment of Res Jost, a Swiss mathematical physicist and close friend who was highly regarded by the mathematicians at the Institute. "I think we could have got him a chair quite easily if Oppenheimer had wanted it. But Oppenheimer was very nasty to him, and Jost was quite bitter when he left here."[47]

Another candidate for whom Dyson pushed unsuccessfully was Jule Charney, a meteorologist who worked close together with John von Neumann as leader of his computational meteorology group. They collaborated on the first numerical weather simulations, initially on the ENIAC, the first programmable computer, at the University of Pennsylvania, and later using the IAS machine that von Neumann had built at the Institute. (These first computations took forty-eight hours to predict tomorrow's weather.) Meteorology was a budding field in which Dyson saw great promise. But after von Neumann's death, the computer and its many applications left the Institute.[48]

Meanwhile, when Oppenheimer vigorously pursued the theoretical physicist C. N. Yang, Dyson pulled in the opposite direction, telling Yang that there would be no opportunities at the Institute. Oppenheimer was furious about this unwanted intervention and gave Dyson "the nastiest dressing-down of my life."[49] But Dyson also had some successes. He was able to bring astrophysics to the Institute with the Danish astronomer Bengt Strömgren, though against Oppenheimer's wishes to make the Institute an exclusive center of particle physics:

I thought we ought to have an astronomer. . . . . Oppenheimer was not at all enthusiastic about that. He had a very narrow view; he thought that anything that was not fundamental science shouldn't be done. And for him, the only thing that was really fundamental was particle physics.[50]

Even in the mid-1960s, when suddenly new opportunities appeared, Oppenheimer pushed through his vision, to Dyson's chagrin:

I said at the time when Yang and [Abraham] Pais and Strömgren all left that the place looked as though it was really going to collapse as far as physics was concerned. And then Oppenheimer quickly made two new appointments, which I wasn't too happy with. Which were Adler and Dashen. . . . . Both of them phenomenologists, and both of them very young at that point. They were rushed through. Oppenheimer was just on the point of retiring and he got these two appointments and nobody opposed them, which surprised me. I guess the mathematicians felt, well, we have to do something to keep the school alive. Anyway, they were approved, which meant the phenomenology continued. And it did, so that all through the seventies and eighties there were lots of phenomenologists here.[51]

Despite all their professional quarrels, Dyson had great feelings of loyalty toward Oppenheimer. When Oppenheimer lost his security clearance in the high-profile 1954 hearing of the Atomic Energy Commission (AEC), it was not at all clear if he could keep his position as director of the Institute. The chair of the AEC, Lewis Strauss, who had personally instigated the security review, also chaired the board of trustees at IAS. In preparation, Dyson had his bags already packed in case Oppenheimer was fired. "I owed my place here to him, and he had my loyalty, no matter what."[52]

Although his own role in the trial was limited to delivering a bag of clean clothes to Oppenheimer through his Washington lawyer, Dyson resented the harsh and deeply unjust treatment by the AEC. He was immensely relieved that the trustees did not follow Strauss and declared their confidence in his leadership of the Institute. Dyson sympathized with Oppenheimer's conflicted and confused loyalties, being "on good terms with the Washington generals and to be a savior of humanity at the same time."[53] In his view, he even became a better director after the tribulations of the trial. "He spent less time in Washington and more time at the institute. He was still a great public figure, a hero to the scientific brotherhood and to the international community of intellectuals, but he became more relaxed and more attentive to our day-to-day problems. He was able to get back to doing what he liked best—reading, thinking and talking about physics."[54]

Their relationship deepened during the later years of Oppenheimer's life. The two grew closer, and Dyson became a close confidant. He started to feel personally responsible for Oppenheimer as his health continued to deteriorate because of an aggressive form of cancer. He was encouraged by Oppenheimer's wife, Kitty:

Kitty believes, perhaps rightly, that I can help Robert to keep alive by keeping alive his interest in physics. She feels desperately that he needs to be convinced that he is still needed in the community of physicists. On the other hand, I find that Robert is so physically tired from the radiation that my instinct is to hold his hand in silence rather than burden him with particles and equations. It is odd that I feel so personally responsible for him. I never had been close to him until now. I suppose it is partly the heredity that runs in our family that makes me want to save souls.[55]

In those final days, Oppenheimer bravely tried to stay engaged with the physics group that still gathered in his office every week to discuss the latest research news:

Oppenheimer didn't want to go to the Cafeteria, and he could hardly eat at that point. So, we had these lunches in his office. I guess it was in D building. After he stopped being director, I think he had this corner room in D building. Anyhow, just a small table. Oppenheimer had some concoction which he drank out of a tall glass. It was just some protein concentrate and that was all he could eat. The rest of us all had sandwiches, which were always very meager. And then we were supposed to discuss what was going on in physics, to keep him up to date. It was really just out of pity for the poor old fellow, to make him feel that he still had some contacts. So, it was a regular occasion. . . . [Marvin] Goldberger and [Sam] Treiman always came from the University, and here we had Adler and Dashen and Regge and myself,

I think. Anyway, about eight of us came to these lunches every Tuesday. We would talk around what was going on in physics. He still wanted desperately to stay in touch. But it was very sad, you saw him fading as the weeks went by.[56]

Oppenheimer passed away on February 18, 1967.

CARL KAYSEN AND THE BELLAH AFFAIR

In 1966, one year before his death, Oppenheimer was succeeded as director by Carl Kaysen. Trained as an economist at Harvard, where he became professor of economics in 1957, his interests shifted to arms control and international relations. As one of the many scholars recruited from Cambridge, Massachusetts, to join the administration of President John F. Kennedy, he was deputy special assistant for national security affairs from 1961 to 1963, second-in-command to national security adviser McGeorge Bundy, his former dean at Harvard.

During the Cuban missile crisis, when the world teetered on the brink of nuclear apocalypse, "Carl was essentially in charge of all other White House foreign policy matters during that time," according to Theodore C. Sorensen, Kennedy's special counsel and speechwriter. "The president had complete confidence in him."[57] Kaysen's leadership at that crucial period of history led some in the White House to nickname him the "vice president in charge of the rest of the world." Kaysen is generally seen as having played a pivotal role in turning the Kennedy administration from the Cold War arms race to an approach of arms control with the Soviet Union.

When the IAS trustees confidentially shared their intention to appoint Kaysen, Dyson's immediate reaction was enthusiastic. He considered him in many ways a "kindred spirit." Indeed, their lives had been remarkably parallel. Kaysen was only three years older. During the Second World War, he served as an American intelligence officer based in the United Kingdom, with the task of helping to pick targets for bombardiers in the Army Air Corps, just as Dyson had done for the Royal Air Force:

While I was receiving my education at Bomber Command from 1943 to 1945, he was just six miles away at the headquarters of the American bomber command at

High Wycombe, doing the same job and receiving the same education. We might even have met on the few occasions when I visited the American HQ.[58]

Kaysen was proud about his humanitarian approach during the war, sending planes to bomb factories and oil refineries instead of civilian areas.

In the service of arms control, they had parallel lives, too. In summer 1963, while Dyson worked at the Disarmament Agency in Washington, DC, Kennedy unexpectedly declared a unilateral moratorium on atmospheric nuclear tests. Kaysen played an important role in the subsequent negotiations of the breakthrough nuclear test ban treaty in Moscow. Although the team was led by W. Averell Harriman, Kaysen was the main actor behind the scenes and kept the president informed daily.[59]

Kaysen had reciprocal feelings of their "semi-parallel activities," writing to Dyson much later in 2006:

I was driving you from Princeton to Cape Cod and we got so involved in exchanging World War Two reminiscences that I forgot to look at the gas gauge. The result: the spectacle of the Director of the Institute for Advanced Study and one of its professors pushing a Volkswagen along the Jersey Turnpike to the fortunately nearby service station.[60]

When Kaysen arrived at the IAS, he expressed his intent to be an active director who wanted the Institute to be fully engaged with contemporary issues, commenting to Dyson, "I did not come here to operate a motel." This was a not-so-subtle reference to an interview that Dyson had recently given, in which he remarked that "the Institute is a motel with stipends":

Like the first director, Abraham Flexner, Kaysen wanted the institute to be active in public affairs. He did not want it to be merely a rest home for scholars. Kaysen's plan was to add a new school to the institute, a School of Social Studies, bringing together a group of people dedicated to understanding, and helping to solve, the problems of modem society.[61]

Indeed, Kaysen was expressly appointed to bring the social sciences to IAS, after an initial effort by Flexner in the 1930s in this direction had failed. He presented his broad vision for social science at the Institute in a fundraising letter:

The goal of all of these studies is an understanding of the forces which shape the direction, pace, and character of change in human societies. The purpose of the Institute's proposed program in this area is to add to fundamental knowledge and understanding rather than to seek to apply this knowledge to problems of economic development or political stability. However, here, as in other areas, it is to be expected that an increase in fundamental knowledge can ultimately contribute to a better understanding of alternative possibilities for action in the real world.[62]

Dyson was immediately supportive of Kaysen and the establishment of social science at the Institute:

I liked him from the start. I don't know how he happened to be contacted. Certainly, I had nothing to do with the initial contacts. He was an unlikely choice. It was fairly obvious that a large group of the faculty didn't want to have social science here and Kaysen made it very clear that he did. So, I don't know. And of course, it was decided by the trustees and not by the faculty. I think at that time the faculty was consulted, but the trustees could appoint anybody they wanted. So, I don't remember too much about it except that I liked him very much and I also liked having social science. And he brought his own funds. So, the social science school didn't use Institute money. It came with its own funds. I forget which foundation it was but anyway, it had enough, and nobody could say that Kaysen had stolen their money.[63]

In 1973, the School of Social Science formally became the fourth school at the Institute. Clifford Geertz, an eminent scholar in the field of cultural anthropology, was the first and founding professor. However, soon the lingering conflicts between faculty and director came to a head. That same year, the sociologist of religion Robert Bellah from the University of California at Berkeley was invited to join the school by Kaysen and Geertz. His candidacy generated strong feelings, particularly by the historians and mathematicians. The opposition was led by the philosopher Morton White, who found Bellah's work "pedestrian and pretentious," and mathematician André Weil, who called Bellah "not of the intellectual and academic quality of a professor at the institute."[64]

At the faculty vote, all the mathematicians and half of the historians voted against the nomination; all the physicists, including Dyson, voted in favor. The result was thirteen to eight with three abstentions. When

Kaysen and the trustees decided to push through the appointment nevertheless, the "divorce" erupted in a public fight. The faculty revolted, and the affair exploded onto the front page of the *New York Times* and other publications, leading to headlines like "Thunderbolts on Olympus," "Ivory Tower Tempest," and "Bad Days on Mount Olympus: The Big Shoot-Out in Princeton."[65]

During all of this, Dyson remained firmly on the side of Kaysen, as he wrote in this letter supporting Bellah's appointment:

What is at stake in the dispute over Bellah's appointment? The basic question is whether a majority of the faculty should wish to impose its standards of personal taste upon a minority belonging to a different field of study. This is the question upon which the trustees have now to pass judgment.

When I first heard of the proposed appointment of Bellah, I read his two books and made up my own mind about them without talking to anybody. Having had negative feelings about both the previous candidates in the Social Science program, I was delighted to find in Bellah's writing an intellectual style to which I could respond with enthusiasm. To me, this was finally the kind of stuff that the social scientists ought to be doing. It is unlikely that even a strongly negative report from the external members of the Ad Hoc Committee would have changed my view. . . .

In conclusion, I urge the trustees to confirm the appointment of Bellah, and to establish once and for all the principle that a majority of the faculty does not have the power, and should not have the wish, to impose its personal tastes upon a minority in a different field.[66]

The affair ended when Bellah retracted his acceptance of the position and returned to Berkeley, also under pressure from the serious illness of his daughter. However, Kaysen still faced harsh backlash jeopardizing his directorship. A committee was formed by the board of trustees that used some strong words to describe the melee but finally decided to support Kaysen:

The Board believes that the conflict of the last several months has deep roots in the history of the Institute, but recognizes that its immediate occasion was the procedure by which Robert Bellah was invited to a professorship in the Program in Social Sciences. Serious dissatisfaction with this procedure by more than half the Faculty has led to a variety of harmful consequences. The first was the airing of the whole dispute in the public press, including an unauthorized leaking of confidential evaluations of Bellah's work. The Board deplores this action which has caused unjustifiable pain to Professor Bellah and done the reputation of the Institute no good.

The same dissatisfaction has led to a request for an evaluation of the Director's tenure, carrying with it a strong expression of a lack of confidence in him by that part of the Faculty. Everything we have heard from the Faculty orally or in writing has led us to conclude that the central feature of that assertion has been his and our procedure in respect to the Bellah appointment. Though Bellah has now withdrawn his acceptance of the appointment, the sharp conflicts it occasioned remain. We reject the view that the appointment procedure justifies a lack of confidence in the fitness of the Director, and his suitability for his post. Rather, we affirm our confidence in him and our recognition of his achievements as Director, which include initiating the Program in Social Sciences, securing of new funding for it, assisting the revivification of the School of Natural Sciences, increasing the funds available to support visiting members, thus making possible a greater diversification of membership in the School of Historical Studies, and organizing the necessary expansion of the Institute's physical facilities.[67]

Subsequently, the revered economist and philosopher Albert O. Hirschman was appointed, a totally uncontroversial matter. The School of Social

Carl Kaysen in 1976. Kaysen served as director of the Institute for Advanced Study between 1966 and 1976. Photograph by Allen Kassof, courtesy of the Shelby White and Leon Levy Archives Center, Institute for Advanced Study, Princeton, New Jersey.

Science soon established itself with further stellar appointments of political theorist Michael Walzer and historian Joan Scott.

These troubles in no way weakened the affection of Dyson for Kaysen; in fact, they made it grow stronger. They remained close friends when Kaysen's directorship finished in 1976 and he moved to MIT as director of the Program in Science, Technology, and Society. One year after he left the Institute, Kaysen wrote to Dyson summarizing his views of the affair:

> Despite the quarrels and pains, it was worth the ten years. I really believe we left the place better than we found it; and I suppose to ask for more, or to expect gratitude, is to seek to transcend the human condition. And that would be impious.[68]

Dyson summarized his relationship with Kaysen as follows:

> Carl was my close friend, and I was fighting by his side for the ten years from 1966 to 1976 that he was Director of the Institute for Advanced Study . . .
>
> As you all know, Carl was in a bad situation at Princeton when he established the School of Social Science at the Institute for Advanced Study. Some Institute mathematicians had never forgiven the Institute Trustees for appointing Robert Oppenheimer as Director twenty years earlier. They decided to take their revenge on Carl. They organized a noisy public campaign against Carl, rather like the campaign of the tea-party Republicans against Obama. . . .
>
> Carl stood firm and ended his ten-year tenure as Director with the Institute peaceful and the School of Social Science running smoothly. During the time of troubles, when insults and threats were flying freely, Carl kept cool. He called the affair the *Froschmäusekrieg*, the war of frogs and mice. He got this word from Helen Dukas, the secretary of Albert Einstein, who got it from Einstein.[69]

Einstein first used the term *Froschmäusekrieg* in 1928 to describe the intense struggle between the German mathematician David Hilbert and the Dutch topologist L. E. J. Brouwer in the editorial board of the prestigious journal *Mathematische Annalen*.[70] The *Battle of the Frogs and Mice* or the *Batrachomyomachia* is a comic parody of the *Iliad* from ancient Greece, most likely written in the Hellenistic period. This "tiny Homeric epic" describes a one-day battle between the mice and frogs. The mice have the upper hand until the Olympian gods send an army of crabs to help the poor frogs. It's an apt depiction of the fight over Bellah—and, for that matter, the intense marital fights between physicists and mathematicians. It reminds one of the famous saying that "academic politics is the most vicious and bitter form of politics, because the stakes are so low."[71]

As we have seen, Dyson loved to describe himself as a frog, immersing himself in the close-up details of a muddy pond before jumping to another exciting adventure. Just like in the *Battle of the Frogs and Mice*, frogs are usually not in the advantage, also in academia, where intellectual transgressions are often punished. However, Dyson could have taken solace from the words of King Pufferthroat in the *Froschmäusekrieg*, praising the advantage of amphibians, equally at home in two worlds—in the case of Dyson, the worlds of mathematics and physics:

> We too have many wonders to explore,
> Both in the pond and the marshy shore,
> For Zeus has given us the power to roam
> In either element, alike at home:
> Amphibian, possessing double lives.[72]

NOTES

1. Freeman Dyson, *Infinite in All Directions* (New York: Harper & Row, 1988), 5.

2. Freeman Dyson, "Birds and frogs," *Notices of the American Mathematical Society* 56, no. 2 (2009): 212–223.

3. Freeman Dyson, letter to his parents, February 1943, reprinted in Freeman Dyson, *Maker of Patterns: An Autobiography through Letters* (New York: Norton, 2018), 29–34, on 33–34.

4. E. J. Kahn and Harold Ross, "Exiles in Princeton," *New Yorker* (10 December 1938): 24–25.

5. Abraham Flexner, *The Usefulness of Useless Knowledge*, rev. ed. (Princeton: Princeton University Press, 2017 [1939]).

6. See David Kaiser, *Drawing Theories Apart: The Dispersion of Feynman Diagrams in Postwar Physics* (Chicago: University of Chicago Press, 2005), 93–98.

7. "The eternal apprentice," *Time* 52 (8 November 1948): 70–81.

8. Lincoln Barnett, "Close up: J. Robert Oppenheimer," *Life* (10 October 1949): 120–138; the 1966 Princeton laudation is quoted in Kai Bird and Martin J. Sherwin, *American Prometheus: The Triumph and Tragedy of J. Robert Oppenheimer* (New York: Knopf, 2006), 582.

9. See, e.g., Kaiser, *Drawing Theories Apart*, 83–93.

10. Richard Feynman, *"Surely You're Joking, Mr. Feynman": Adventures of a Curious Character* (New York: Bantam Books, 1986), 149.

11. Stewart quoted in Bird and Sherwin, *American Prometheus*, 387.

12. The prize citation is available at https://www.nobelprize.org/prizes/physics /1957/summary/.

13. "Institute for Advanced Study celebrates Freeman Dyson at 90," accessed 13 April 2021 at https://www.ias.edu/press-releases/institute-advanced-study -celebrates-freeman-dyson-90.

14. For an accessible account, see Robert Crease and Charles Mann, *The Second Creation: Makers of the Revolution in Twentieth-Century Physics* (New York: Macmillan, 1986).

15. Kaiser, *Drawing Theories Apart*, 320.

16. Feynman quoted in Joel E. Cohen, "A life of the immeasurable mind," *Annals of Probability* 14, no. 4 (1986): 1139–1148, on 1147.

17. Peter Galison, "Structure of crystal, bucket of dust," in *Circles Disturbed: The Interplay of Mathematics and Narrative*, ed. Apostolos Doxiadis and Barry Mazur (Princeton: Princeton University Press, 2012), 52–78, on 65.

18. Liliane Beaulieu, "A Parisian café and ten proto-Bourbaki meetings (1934– 1935)," *Mathematical Intelligencer* 15, no. 1 (1993): 27–35; Leo Corry, "Nicholas Bourbaki: Theory of structures," in Corry, *Modern Algebra and the Rise of Mathematical Structures* (New York: Springer, 2004), 289–338.

19. André Weil, *The Apprenticeship of a Mathematician* (Boston: Birkhäuser, 1992), 114.

20. Silvan Schweber, "The empiricist temper regnant: Theoretical physics in the United States, 1920–1950," *Historical Studies in the Physical and Biological Sciences* 17 (1986): 55–98.

21. Natalie Wolchover interview with Freeman Dyson and Karen Uhlenbeck, The Universe Speaks in Numbers Symposium, Institute for Advanced Study (29 May 2019), www.youtube.com/watch?v=jZBtuycmAbE (Dyson comments at 03:20).

22. Freeman Dyson, interview by George Dyson, 1–5 May 2004, 10.

23. Freeman Dyson, "Missed opportunities," *Bulletin of the American Mathematical Society* 78, no. 5 (1972): 635–652, on 635.

24. J. Hadamard, *The Psychology of Invention in the Mathematical Field* (Princeton: Princeton University Press, 1945), on 54.

25. Despite the emphasis on particle physics at IAS in the 1950s, many crucial contributions to the field of statistical mechanics were made at that time too, for example, the seminal work of T. D. Lee and C. N. Yang in 1952 on phase transitions of large physical systems.

26. Freeman Dyson, "Ground state energy of a hard-sphere gas," *Physical Review* 106 (1957): 20–26; Freeman Dyson, *Selected Papers with Commentary* (Providence, RI: American Mathematical Society, 1996), 23 ("do better things").

27. A classic discussion is Alexandre Koyré, *From the Closed World to the Infinite Universe*, 2nd ed. (Baltimore: Johns Hopkins University Press, 1968 [1957]); see also Roger Hahn, *Pierre Simon Laplace, 1749–1827: A Determined Scientist* (Cambridge, MA: Harvard University Press, 2005).

28. For an accessible account, see David Ruelle, *Chance and Chaos* (Princeton: Princeton University Press, 1991).

29. James H. Jeans, *The Mathematical Theory of Electricity and Magnetism* (Cambridge: Cambridge University Press, 1908).

30. Paul Ehrenfest's address on the occasion of the awarding of the Lorentz Medal to Wolfgang Pauli on 31 October 1931, translated and reprinted in M. J. Klein, ed., *Paul Ehrenfest: Collected Scientific Papers* (Amsterdam: North-Holland, 1959), 617.

31. Dyson, *Selected Papers*, 32.

32. Freeman Dyson, "Ground-state energy of a finite system of charged particles," *Journal of Mathematical Physics* 8 (1967): 1538–1545.

33. Dyson, *Selected Papers*, 32–33; see also Elliott H. Lieb and Robert Seiringer, *Stability of Matter in Quantum Mechanics* (Cambridge: Cambridge University Press, 2010).

34. Madan Lal Mehta, *Random Matrices*, 3rd ed. (New York: Academic Press, 2004 [1967].

35. Freeman Dyson, interview by George Dyson, 1–5 May 2004, 67.

36. Freeman Dyson, "The threefold way: Algebraic structure of symmetry groups and ensembles in quantum mechanics," *Journal of Mathematical Physics* 3 (1962): 1199–1215.

37. Dyson describes his paper on "the threefold way" as his "favorite" in Dyson, *Selected Papers*, 28.

38. One should mention here the beautiful Dyson process that describes the evolution of the eigenvalues in time in terms of random processes.

39. Freeman Dyson, interview by George Dyson, 1–5 May 2004, 67.

40. Albert Einstein, Boris Podolsky, and Nathan Rosen, "Can quantum-mechanical description of reality be considered complete?," *Physical Review* 47 (1935): 777–780; see also Louisa Gilder, *The Age of Entanglement: When Quantum Physics Was Reborn* (New York: Knopf, 2008).

41. For accessible introductions, see Dan Rockmore, *Stalking the Riemann Hypothesis: The Quest to Find the Hidden Law of Prime Numbers* (New York: Pantheon, 2005); and Alex Kontorovich, "How I learned to love and fear the Riemann hypothesis," *Quanta* (4 January 2021), https://www.quantamagazine.org/how-i-learned-to-love-and-fear-the-riemann-hypothesis-20210104/.

42. Wolchover interview with Dyson and Uhlenbeck, 29 May 2019, https://www.youtube.com/watch?v=jZBtuycmAbE.

43. Graham Farmelo, "Remembering Freeman" (2020), https://grahamfarmelo.com/remembering-freeman

44. Freeman Dyson, podcast interview with Graham Farmelo, "The Universe Speaks in Numbers" symposium (May 2019), https://grahamfarmelo.com/the-universe-speaks-in-numbers-interview-10/ (quotation at 07:56).

45. Freeman Dyson, interview by Silvan Schweber, on Web of Stories, 1998, https://www.webofstories.com/play/freeman.dyson/83.

46. Freeman Dyson, "Oppenheimer: The shape of genius," *New York Review of Books* (15 August 2003).

47. Freeman Dyson, interview by George Dyson, 1–5 May 2004, 65.

48. George Dyson, *Turing's Cathedral: The Origins of the Digital Universe* (New York: Pantheon, 2012).

49. Graham Farmelo, "Remembering Freeman" (2020), https://grahamfarmelo.com/remembering-freeman.

50. Freeman Dyson, interview by George Dyson, 1–5 May 2004, 21.

51. Freeman Dyson, interview by George Dyson, 1–5 May 2004, 66.

52. Graham Farmelo, "Remembering Freeman" (2020), https://grahamfarmelo .com/remembering-freeman. On Oppenheimer's 1954 hearing, see Bird and Sherwin, *American Prometheus*, chaps. 34–38.

53. Freeman Dyson, *Disturbing the Universe* (New York: Harper & Row, 1979), 87.

54. Freeman Dyson, *Disturbing the Universe*, 76.

55. Freeman Dyson to his parents, 30 March 1966, reprinted in Dyson, *Maker of Patterns*, 321.

56. Freeman Dyson, interview by George Dyson, 1–5 May 2004, 71.

57. Bryan Marquand, "Carl Kaysen, 89, MIT professor, economist, and JFK advisor," *Boston Globe* (9 February 2010).

58. Freeman Dyson to his parents, 17 February 1966, reprinted in Dyson, *Maker of Patterns*, 319.

59. Dennis Hevesi, "Carl Kaysen, nuclear test ban negotiator (and much more), dies at 89," *New York Times* (19 February 2010).

60. Carl Kaysen, letter to Freeman Dyson, 9 November 2006, quoted in Freeman Dyson, "Reflections on a friendship with Carl Kaysen," *Institute Letter* (Summer 2010).

61. Dyson, *Maker of Patterns*, 322. Early in his tenure as IAS director, Oppenheimer had explained that his goal was to transform the Institute into "an intellectual hotel": quoted in "Eternal apprentice," *Time* (8 November 1948), 81.

62. Fundraising Letter to Ford Foundation, "School of Social Science Fund (1968)," in the Shelby White and Leon Levy Archives Center, Institute for Advanced Study, Princeton, NJ.

63. Freeman Dyson, interview by George Dyson, 1–5 May 2004, 46.

64. These and similar critiques were reported in the front-page article: Israel Shenker, "Dispute splits Advanced Study Institute," *New York Times* (2 March 1973).

65. J. K. Footlick, "Thunderbolts on Olympus," *Newsweek* (19 March 1973): 60–63; "Ivory tower tempest," *Time Magazine* (19 March 1973): 48; L. Y. Jones Jr., "Bad days on Mount Olympus: The big shoot-out at the Institute for Advanced Study," *Atlantic Monthly* 233 (February 1974): 37–53. See also Matteo Bortolini, "The 'Bellah

Affair' at Princeton: Scholarly excellence and academic freedom in America in the 1970s," *American Sociologist* 42 (2011): 3–33.

66. Board of Trustees Records, Committee Files, box 1: Minutes—Executive Committee, BOT 12-20-72; letter from Freeman Dyson to Carl Kaysen in support of Robert Bellah appointment, 50–51, in the Shelby White and Leon Levy Archives Center, Institute for Advanced Study, Princeton, NJ.

67. Board of Trustees Records, Committee Files, box 3: Board Memo on Dilworth Committee to Faculty, 27 April 1973, in the Shelby White and Leon Levy Archives Center, Institute for Advanced Study, Princeton, NJ.

68. Freeman Dyson, remarks at a memorial for Carl Kaysen held at the American Academy of Arts and Sciences, 22 May 2010, https://www.ias.edu/ideas/2010/dyson-kaysen.

69. Dyson, remarks at a memorial for Carl Kaysen.

70. *The Collected Papers of Albert Einstein*, vol. 16: *The Berlin Years / Writings and Correspondence/June 1927–May 1929* (Princeton: Princeton University Press, 2021), lxxviii.

71. Political scientist Wallace Stanley Sayre, quoted in *Wall Street Journal* (20 December 1973).

72. A. E. Stallings, *The Battle between the Frogs and the Mice: A Tiny Homeric Epic* (Philadelphia: Paul Dry Books, 2019).

5   SINGLE STAGE TO SATURN: PROJECT ORION, 1957–1965

GEORGE DYSON

The question of how to go into space cannot be separated from the question of why.
—Freeman Dyson, "Mankind in the Universe" (1969)

"The first rocket mail service in England was a big event," Freeman Dyson remembered. "The British post office needed to take mail from the mainland to the Isle of Wight and it took half a day to get mail on the ferry and take it across. Someone sold them on the idea of sending this mail across on rockets. It was just pure luck that we lived very close to the launch site. So we watched them launch."[1]

In 1930 Dyson's father had adopted an old farmhouse on forty waterlogged acres near Lymington, some thirty miles southwest of Winchester, as a retreat from urban life. The silence was conducive to his work as a composer, and the absence of modern conveniences exposed his children to some of the manual labor they were otherwise spared. On 19 December 1934, just four days after Freeman's eleventh birthday, a rocket launch was a welcome escape from clearing the drains and tending the dikes.

"They had this impressive looking rocket and set it up on this derelict piece of land where we lived which was sort of a mud flat on the coast opposite the Isle of Wight," Dyson explained. "Some dignitaries from London came down and ceremoniously put this bag of mail with special stamps into the rocket. And it zoomed up into the sky very beautifully. But then it turned around and came back almost exactly where it took off and landed with a big splash in the mud. So they went out and retrieved it, and the mail went over later on the boat."[2] The venture was organized by a German promoter named Gerhard Zucker, whose business model, based on selling rocket mail stamps to collectors, paid off whether rocket delivery was successful or not.

Ten years later, while serving in Operations Research at Royal Air Force Bomber Command, Dyson bore witness to a venture organized by Zucker's compatriot Wernher von Braun. It was also a mixed success. "We were very grateful to Wernher von Braun," Dyson explained. "We knew that each V-2 rocket cost as much to produce as a high-performance fighter aircraft."[3] German fighters were inflicting heavy and repeated losses on the Allied bomber squadrons, whereas each V-2 killed an average of about two people at its target and only exploded once.

When V-2s began falling on London, Dyson's thoughts turned to space. "I remember being delighted to learn that the V-2 really existed. It was a big step forward. It went 50 miles up and 250 miles horizontal. If you could do that much you could get into space. And then I was rather disappointed. If the Germans could do that well, I expected that we would have our own secret projects. Probably we would be doing much better. At the end of the war I found out there was really nothing on our side. We had to start over again from scratch."[4]

After the loss of too many scientists in the First World War, Britain had decided to spare them from combat in the Second World War. Dyson, assigned to statistician Reuben Smeed, worked among the bomber crews, who faced a one-in-twenty chance of death on every mission, but except for one or two test flights, he was kept on the ground. It was a safe but depressing existence: calculating alone while boys his own age and younger risked their lives in tightly knit groups. His analyses revealed many of the assumptions underlying the bombing strategy to be wrong, but this information was suppressed as it filtered back up the chain of command.

The scientists assigned to Los Alamos, in contrast, were sequestered with their families and given both a clearly defined objective and the resources to carry it out. "They did not just build the bomb," Dyson later explained. "They enjoyed building it. They had the best time of their lives building it. That, I believe, is what Oppenheimer had in mind when he said that they had sinned."[5]

When Dyson arrived in America in 1947, the excitement of Los Alamos was over, but Hans Bethe's group was keeping its spirit alive at Cornell. Dyson's introduction to laboratory work ended badly, with the high voltage intended to energize an electric field in the Millikan oil drop experiment passing through him instead. A life sentence to theoretical physics loomed over his head.

GENERAL ATOMIC

Dyson remained a theorist until Frederic (Freddy) de Hoffman, a protégé of Edward Teller and a fellow graduate student under Bethe at Cornell, went into the atomic energy business and offered him a summer job out west. When De Hoffmann was twenty years old and working at Los Alamos, he had calculated the ballistic trajectories for the atomic bombs that were dropped on Japan. "Making the bomb tables for Hiroshima and Nagasaki is etched so strongly in my mind because it really brought me to the reality of the end use of our scientific and technical experimentation," he explained.[6] He also performed the initial calculations for what was then known as the "Super" bomb. "I wanted to do something about the hydrogen bomb and nobody else wanted to," remembered Teller. "And the one man who wanted to do it more than I was Freddy de Hoffmann."[7]

After leaving Los Alamos and completing his PhD, de Hoffmann became chair of the Committee of Senior Reviewers of the Atomic Energy Commission (AEC) and was appointed a scientific secretary to the Atoms for Peace conference convened by the United Nations in Geneva in summer

1955. Commercial applications of atomic energy were about to be declassified, and John Jay Hopkins, chairman of General Dynamics, the largest US defense contractor, asked Edward Teller for advice. Teller answered: "Freddy de Hoffmann." Hopkins gave de Hoffmann $10 million to start things up.

"De Hoffmann was Napoleonic by nature," explained Brian Dunne, an experimentalist who joined de Hoffman's group. "He studied Napoleon's life as a boy. He had this dream, this vision, but it was driven by grandiosity. Those kinds of fevers are catching things."[8] General Atomic (now General Atomics) was founded on 18 July 1955, with the mission, as de Hoffmann put it, to "bring the sun down to earth." A year later, the company was operating out of a converted elementary school near Point Loma in San Diego, vacant since the exodus of military families after the war. "The drinking fountains were down very low for children and the blackboards were low," remembered Dunne. "The machine shop used to be the kindergarten and all the drawers were way down there next to the floor."[9]

General Atomic's main objective was fusion power for peaceful purposes, going by the code name Project Sherwood, since anything concerning fusion was still classified at that time. As a pilot project, de Hoffmann decided to build an intrinsically safe fission reactor, designed to shut down in milliseconds, without human or mechanical intervention, if the control rods were removed or cooling failed. The goal, as Edward Teller put it, was a reactor that was "not only idiot-proof, but Ph.D.-proof." Dyson, whose clearance to work on fusion had been delayed due to his British citizenship, jumped in to help. "I am amusing myself with uranium reactors and I find it absorbingly interesting," he reported to his parents. "Probably this summer is a turning-point in my life. I find the atomic energy business not only congenial, but also find I am good at it. My real talent is perhaps not so much in pure science as in the practical development of it."[10]

Dyson distilled the theory behind the safe reactor into a four-page paper, issued as the company's sixth technical report.[11] "If I hadn't had [Dyson's] little paper I couldn't have made the guesses to make those experiments," said Brian Dunne, who helped to get the construction of the reactor off the ground. "It was like the first draft of what you'd put on a patent application—he gave guesses as to the total amount of U238, U235, zirconium, hydrogen . . . I can trust this as the basis of the design for this thing."[12] The physicists gathered at the schoolhouse recaptured the enthusiasm of wartime Los Alamos, transgressing the boundaries between physics and engineering to advance from a theory of the warm neutron effect to a two-megawatt prototype in under two years. When the second United Nations Atoms for Peace conference convened in Geneva in September 1958, General Atomic showed up with a working reactor and stole the show. "Everyone wanted to see the blue light," recalled Dunne. "They sold those things like hotcakes."[13]

Sixty-six copies of the TRIGA (Training, Research, Isotopes, General Atomics) reactor were commissioned, some still in operation today. TRIGA was consistently profitable, a record unmatched by any other nonmilitary reactor design. Dyson left his name, along with Andrew W. McReynolds and Theodore B. Taylor, on a patent for a "Reactor with Prompt Negative Temperature Coefficient and Fuel Element Therefor." They received one dollar each for their rights. "The primary object of the present invention is to provide an improved neutronic reactor which will not be destroyed even if grossly mishandled," they explained.[14]

De Hoffmann believed the reactor's safety needed to be demonstrated in a spectacular way. At the public dedication, with Niels Bohr in attendance, "we pulled out all the control rods explosively so that the reactor was prompt super-critical with a neutron doubling time of two milliseconds," Dyson remembered. "That was the worst accident that we could imagine. The reactor quietly shut itself down in a few milliseconds with no damage. We then repeated the process five hundred times."[15]

Q CLEARANCE

In August 1956, there was a sudden panic over a shortage of tritium for nuclear weapons, and Dyson was diverted from the TRIGA project under a temporary emergency clearance to help. "The answer came back

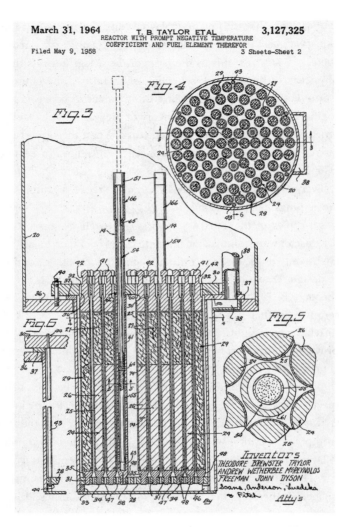

Figures from US Patent no. 3,127,325 for the TRIGA reactor, filed 9 May 1958 by Theodore Taylor, Andrew McReynolds, and Freeman Dyson. Source: US Patent and Trademark Office.

from the security people that it would not be possible to grant a clearance to Mr. Dyson for any kind of civilian application of atomic energy," he remembered. "On the other hand, exceptions could be made in the case of projects which were of urgent relevance to national security so therefore I would be cleared for the military stuff. It made no sense."[16]

On 19 September Dyson received his unrestricted Q clearance from the authorities in Washington and flew to Los Alamos the next day. "To my amazement they simply stuffed me with all their information about bombs," he recalled. "I hadn't asked for that, I wasn't particularly interested in bombs. They wanted to tell me everything they'd been doing, as if they'd just sort of been burning to talk about this to somebody, all the designs that they had done and what they were planning to do. So I listened to all this, I didn't do anything, and came back here to Princeton and resumed the normal life."[17] The tritium panic, code-named Project August, turned out to be a false alarm, but the Q clearance became Dyson's ticket to fifteen months that he called "the most exciting and in many ways the happiest of my scientific life."[18]

Theodore "Ted" Taylor, Dyson's collaborator on both TRIGA and Project Orion, was a young physicist from Los Alamos who had designed the most powerful fission weapons ever to populate the US stockpile before he obtained his PhD. He had grown up in Mexico City, making his own high explosives and prowling the streets of Mexico City, looking for places "where the billiard balls were really spherical and the tables were very heavy and flat."[19]

After joining the navy during the Second World War, Taylor obtained a degree in physics from Caltech through the V-12 scholarship program before heading to Berkeley for a PhD, where he failed his preliminary examinations, twice. In February 1949, newly married and with a child on the way, he was running out of options when Carson Mark, the Canadian physicist who had succeeded Hans Bethe as director of T-Division at Los Alamos, offered him a job working on "problems in neutron diffusion," paying $375 a month—more than twice the typical rate for graduate students' stipends at the time. Taylor's insights into nuclear explosions manifested in an uncanny, intuitive way. It was a game of billiards played with atomic nuclei, and at Los Alamos, the tables were perfectly flat.

He was soon designing bombs whose yields were larger than had been thought possible while making them smaller and more efficient as well. The largest fission weapon that Taylor designed was the Super Oralloy Bomb (SOB), which yielded five hundred kilotons when tested at Eniwetok Island in November 1952. The smallest remains unknown, although he admitted that he explored the possibility of "getting the quantities of plutonium to make nuclear explosions down into less than a kilogram. Quite a bit less. I tried to find out what was the smallest bomb you could produce. It was a full implosion bomb that you could hold in one hand that was about six inches in diameter."[20]

In 1953, Taylor was sent on paid leave to Cornell, under Hans Bethe's supervision, to obtain his elusive PhD. It had become a growing embarrassment to have the design of the nation's most advanced weapons in the hands of someone without an advanced degree. Dyson and Taylor, who met casually at Cornell, became colleagues during the summer at General Atomic in 1956. At the end of the summer, Taylor accepted an offer to remain at General Atomic, finding life in La Jolla more suited to his growing family than Los Alamos. He was ready to settle down to designing reactors instead of bombs. Then *Sputnik* went up.

SPUTNIK

On October 4, 1957, the Soviet Union launched an artificial satellite into orbit around Earth. What could the United States launch in response? "That night," Taylor recalled, "I derived for myself the notion that if you really added together the features that you wanted of any vehicles for exploring the solar system—the whole thing, not just near in—you're led directly to energy on the scale of a lot of nuclear weapons. And having been led to that, in thinking about what might be done, I began saying, 'Gee, that's what Stan Ulam's been talking about all these years.'"[21]

The Polish mathematician Stanislaw Ulam, one of Taylor's mentors at Los Alamos, had observed that parts of the Trinity shot tower, made of ordinary steel, had not been vaporized by the fireball and had survived the explosion intact. He began wondering: why not use nuclear explosions to propel rockets rather than use rockets to deliver bombs? He pushed the idea as far as to take out a patent, with Cornelius J. Everett, for a "nuclear propelled vehicle, such as a rocket," and wrote a Los Alamos report, "On a Method of Propulsion of Projectiles by Means of External Nuclear Explosions," issued (in secret) in August 1955.[22]

"Repeated nuclear explosions outside the body of a projectile are considered as providing a means to accelerate such objects to velocities of the order of 10^6 cm/sec . . . in the range of the missiles considered for intercontinental warfare and even more perhaps, for escape from the earth's gravitational field," they explained. The energy and temperature limitations governing internal combustion rockets could be raised by several orders of magnitude by taking a nuclear reactor to its extreme—the explosion of a bomb—and isolating the resulting high temperatures from the ship. "The scheme proposed in the present report involves the use of a series of expendable reactors (fission bombs) ejected and detonated at a considerable distance from the vehicle, which liberate the required energy in an external 'motor' consisting essentially of empty space."[23]

Can you blow something *up* without blowing it apart? The answer appeared to be yes. "I was up all night and then I got alarmed that things were getting big," Taylor continued:

Energy divided by volume is giving pressure, so the pressures were out of sight, unless it was very big. It got easier as it got bigger. I was thinking of something that might carry a couple people, with shock absorbers. I went in to General Atomic the next morning and my office was right next to Chuck Loomis, who had come down from Los Alamos, and I told him about this sense of discouragement because it was so big. And he said, 'Well, think big! If it isn't big, it's the wrong concept. What's wrong with it being big?' In less than thirty seconds everything flipped. It was Chuck's call that if you were serious about exploring the solar system, why not use something the size of the *Queen Mary*? He understood that bombs could in principle do it. They could lift downtown Chicago into orbit.[24]

In response to *Sputnik*, President Eisenhower established the Advanced Research Projects Agency (then ARPA, now DARPA) to coordinate existing civilian and military space programs in the effort to catch up. NASA, requiring an act of Congress, would not exist until July 1958. There was a brief window of opportunity where a project using military weapons for a nonmilitary mission could find a home.

General Atomic issued Taylor's "Note on the Possibility of Nuclear Propulsion of a Very Large Vehicle at Greater Than Earth Escape Velocities" on 3 November 1957, the day that *Sputnik II*, carrying the dog Laika, was launched. The new project was code-named Orion—for no particular reason, remembered Taylor, who just "picked the name out of the sky." At the end of November, de Hoffmann went to Princeton to recruit Dyson, the proposal being too secret to discuss by mail or phone. "He came here and he said, 'Look, you've got to come to GA [General Atomic],'" Dyson recalled. "I said no, I have no intention of coming to GA, I've done my bit for GA. And he said no, you must come, we have something much bigger and much more exciting, and then he told me Ted had this wonderful scheme for getting around the solar system with bombs."[25]

ARPA

On New Year's Day 1958, Dyson flew from Philadelphia to San Diego to consult with Taylor for ten days. "The miserable state of American rockets at present is only to be compared to the state of the English weapons development in 1939," he wrote to his parents on the long flight west, adding, "I expect eventually to take a hand in this myself."[26] He spent his days working on the spaceship and the evenings completing a paper on mathematical aspects of quantum field theory for the *Physical Review*. "I was never happier and more free from worries about anything," he reported on the trip home.[27] The resulting proposal, submitted to ARPA in early 1958, "included all the necessary practical working features for a very large space vehicle" weighing four thousand tons, carrying twenty-six hundred bombs, and capable of orbiting a payload of sixteen hundred tons.[28]

Dyson had "derived certain engineering parameters from physical first principles in an amazingly clear way," remembered then-Major, and later General, Lew Allen, assigned to review the proposal for the Air Force Office of Special Projects. "He ended up showing that really the only number that mattered was the strength of materials. Once the strength of materials was set—as it was for, say, steel—all the other parameters of the vehicle naturally fell out. It was such a beautiful and simple way to look at that."[29] Another Air Force reviewer noted that "the uses for ORION appeared as limitless as space itself."[30]

When physicist Herbert F. York was appointed technical director of ARPA in March 1958, he found a proposal for a four-thousand-ton interplanetary spaceship on his desk. "It was the time after *Sputnik* when everybody was looking for some kind of an answer and thinking that technology was the likeliest place to look, and so a lot of stuff that would be too far out under ordinary circumstances managed to get included inside the envelope," he explained.[31] York was skeptical of estimated costs. "It wasn't just Orion," he said. "Almost every cost estimate made by a physicist is wildly wrong, and the better the physicist the worse it is."[32]

York authorized a one-year feasibility study with "the verbal understanding that the contract would be extended at a somewhat higher rate if it proved technically impossible to disprove feasibility at the end of the first year." He believed the idea deserved a chance. "It was a unique time. When we were getting ARPA started we were willing to take some fliers. I never thought it was feasible, but that's okay, I thought it was interesting. And of sufficiently dramatic ultimate potential that even very low feasibility merited some attention. I tried to put that combination together somehow and multiply that out."[33]

ARPA's contract for the Feasibility Study of a Nuclear Bomb Propelled Space Vehicle was issued on 30 June 1958, between the General Atomic division of General Dynamics Corporation and the US Air Force Air Research and Development Command. "If the concept is feasible, it may be possible to propel a vehicle weighing several thousand tons to velocities several times earth escape velocities," the contract noted. "Such a vehicle would represent a major advance."[34]

The amount awarded was $949,550 plus a fixed fee of $50,200 for a total of $999,750. "There must have been a million-dollar limit," recalled Ed Giller, a young physicist and Air Force colonel who was assigned management of the contract on the Air Force side. "Right up against the peg!" added his colleague Don Prickett.[35]

PAUL AND BARNABAS

In early spring 1958, General Atomic began moving to a futuristic new laboratory just north of La Jolla at Torrey Pines. The campus was laid out around a circular technical library with a courtyard in the middle and a cafeteria on the upper floor. The building, elevated on steel columns around its periphery that could stand in for shock absorbers, was 135 feet in diameter: the exact dimensions of the four-thousand-ton Orion design. Taylor would point to a car or a delivery truck, the size of existing space vehicles, and say, "This is the one for looking through the keyhole." Then he would point to the library and say, "And this is the one for opening the door."[36]

The project was kept entirely secret until May 1958 when General Atomic received permission to reveal its existence for recruiting purposes, and on 2 July 1958, a one-page press release announced "the possible development of a new concept of propulsion employing controlled nuclear explosions . . . within the atmosphere and beyond."[37] Over the July Fourth holiday, Dyson sent a five-page letter to Oppenheimer asking for a leave of absence from the Institute for Advanced Study to stay in La Jolla and join the project full time. "Either through a misunderstanding of our intentions, or by an unusual and deliberate act of wisdom, the government has announced to the public that we are working on the design of a spaceship to be driven by atomic bombs," he wrote. "This scheme alone, of all the many space-ship schemes which are under development, can lead to a ship adequate to the real magnitude of the task of exploring the Solar System."

He added a statement of principles:

1) There are more things in heaven and earth than are dreamed of in our present-day science. And we shall only find out what they are if we go out and look for them.

2) It is in the long run essential to the growth of any new high civilization that small groups of men can escape from their governments and go and live as they please in the wilderness. Such isolation is no longer possible on this planet.

3) We have for the first time found a way to use the huge stockpiles of our bombs to better purpose than for murdering people.[38]

Oppenheimer approved the request. Dyson now became a full-time technologist for the first and only time in his life. In Project Orion, he found the mission he had always dreamed of, and in Ted Taylor, he found the brother he never had. "Ted's genius was the courage that led him to question the obvious impossibility that one can sit on top of a bomb without being fried," he reported to his parents. "He is the Columbus of the new age. It will work, and it will open the skies to us. . . . Ted and I will fly together to Los Alamos this evening. We travel like Paul and Barnabas. Golly, this life is good."[39]

HOTTER THAN THE SUN, COOLER THAN A BOMB

Because there was no existing brotherhood of nuclear-bomb-propelled spaceship designers to defend the territory, Dyson and Taylor had the field to themselves. "In the course of his deciding whether to stay with the project and resign from the Institute, [Dyson] said he had to make a choice between being a very good theoretical physicist or the best engineer ever," Taylor recalled.[40] The landscape was unexplored, starting with how to convert the energy of a nuclear explosion into momentum that could drive a vehicle and ending with where to go in the spaceship and how much fallout would be left behind.

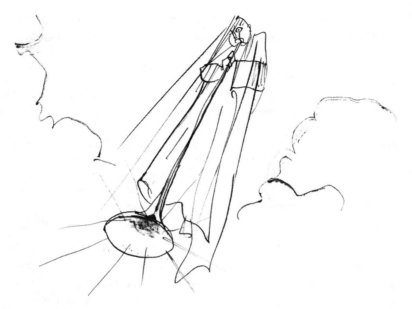

"Either through a misunderstanding of our intentions, or by an unusual and deliberate act of wisdom, the government has announced to the public that we are working on the design of a space-ship to be driven by atomic bombs."

What happens when a solid object in a vacuum is transformed, within a few millionths of a second, into a very hot gas? In three reports, collectively titled *Free Expansion of a Gas*, Dyson addressed this question, warning that "a real cloud of gas will not have precisely the density-distribution of the model, but still one may expect the behavior of a real cloud to be qualitatively similar to that of the model." When the numbers were run on an IBM 650 punched-card programmable calculator, the workhorse of its time, it turned out, as expected, that "the final ellipsoid is prolate when the initial ellipsoid is oblate, and vice versa."[41]

"You get inversion," Dyson explained. "Something that begins like a pancake becomes like a cigar, and something that begins as a cigar becomes a pancake." This was the first of a series of numerical results that fell out in Orion's favor, allowing the project to move ahead. "You start out with a pancake and it will produce then a jet which is collimated within 20 degrees or so quite nicely. The fact that it's so easy to make an asymmetrical explosion may still be classified, for all I know."[42] The thinner the pancake, the narrower the jet.

The development of hydrogen bombs in the early 1950s depended on understanding, and tailoring, the radiation transport and radiation hydrodynamics that channeled the energy from the primary fission explosion so that the thermonuclear fuel was compressed sufficiently to ignite. The transfer of energy to Orion's pusher plate via the pancake of propellant was analogous to the first stage of a hydrogen bomb turned inside out.

The initial explosion compresses the propellant slab, which then expands as a jet of plasma moving at some 150 km/sec (300,000 mph) toward the ship. It takes about 300 microseconds to make the trip. During this time the expanding propellant cools to about 10,000 degrees (Kelvin), until it hits the pusher (or the advancing front of a reflected shock wave) and is suddenly recompressed. For less than a millisecond the stagnating propellant reaches a temperature of between 100,000 and 120,000 degrees—hotter than the surface of the sun but cooler than a bomb.

In making its way through the atmosphere, the ship would ascend upon a series of fireballs, but once in space, "the debris goes out from the bomb essentially invisibly," Dyson explained. "You don't see anything until the stuff is stopped. Around the bomb you have a lot of cold stuff which absorbs the energy so the debris comes out forwards and backwards and that won't produce anything very spectacular in the way of a flash until it hits the ship. Then all its energy is converted into heat and so you get about a millisecond or so of intense white flash. And very little else."[43]

Orion's feasibility depends on what happens as the hot plasma stagnates against the plate. After two or three thousand impacts, would there be any spaceship left? The time is too short for heat to penetrate the surface, so the plate is able to survive an extended series of pulses without melting, the way someone can run barefoot across a bed of coals. Even on an interplanetary mission, involving several thousand explosions, the total plasma-pusher interaction time amounts to less than one second. The question was how much the surface would ablate.

"The most important thing is how opaque the stuff is," Dyson explained. "This whole business of opacity is the central problem both in stars and in bombs. The opacity is like the resistivity of a metal except you are dealing with radiation instead of electrons. It tells you how hard it is for the radiation to get through."[44] Opacity is where Orion either succeeds or fails. "It just kept coming up," added Taylor. "Opacity was repeated ten times a day."[45] If the material is opaque, it prevents harmful radiation from reaching the surface of the pusher. It also blocks secondary radiation produced by the collision between the plasma and the pusher from escaping back through the layer of stagnating propellant, thus doubling the kick. "If it is sufficiently opaque so you don't lose that energy by radiation, then it bounces back and you get the momentum doubled. If it is transparent, the heat radiates away and you lose it; it is just like a slab of mud being thrown against something and you get the momentum it had originally, nothing more."[46]

Nature appeared to be on the side of Orion. "If you have, roughly speaking, a bomb which is a hundred meters away from the ship with a yield of a kiloton, the temperature works out at a hundred thousand degrees," Dyson explained. "This was an unusual temperature which had never been thought about much because stars are generally cooler and bombs are usually hotter. So this was an intermediate range."[47] Opacity increases as the plasma piles up. "The densities we were talking about were, roughly speaking, one gram per liter, or normal air density, which is unusual for something that hot. The more dense it is the more opaque it gets; if you squeeze the stuff together it gets blacker. Nobody had calculated this before."[48] Opacities had been studied intensively and in secret at Los Alamos, Livermore, and RAND. The opacity of heavy elements like uranium at high temperatures was essential to hydrogen bomb design, and the opacity of air at lower temperatures was critical to understanding fireball development and nuclear weapons effects. Orion fell in between.

A Note on Maximum Opacity was Dyson's first nonclassified Project Orion report.[49] "We started doing a much better job than anyone had done before, doing it atom by atom, not just using averages," said Dyson:

These atoms all have very complicated spectra and everything depends on windows because it's where the atom doesn't absorb that the radiation gets through. The important thing is to get the exact shape of the windows right. It's a delicate calculation. And to fill in the windows it's important to have a mixture of things: carbon and nitrogen and oxygen which have windows in different places so they fill in each other's windows. And you need the hydrogen just in order to have the chemical compounds that are easy to handle, like polyethylene, which is good stuff physically and also reasonably opaque. You prefer to have something with nitrogen and oxygen as well. But we generally thought of polyethylene as being good enough. It turned out the opacities were pretty high, even just for carbon by itself. The results always turned out to be rather good from the point of view of feasibility.[50]

The next question was what happens at a surface being ablated not only by a direct impact but also by a lateral flow as the incoming plasma begins to mix with material being evaporated from the surface of the plate. "The question is, when is that stable and when is it unstable," said Dyson. "The answer was that it was generally stable, but you couldn't be sure."[51] Convection and turbulence between the layers of stagnating propellant and the ablating surface might defeat the self-protection of the pusher. "I did a calculation looking at the worst case," Dyson recalled, whose *Preliminary Study of Convective Ablation* reported the results.[52] "If the thing was totally unstable and convective, then how bad would the ablation be? And it turned out even in that case it wasn't terribly bad. Because the time is so short, convection only has time to go around once or twice, so even in the worst case the stuff doesn't ablate more than is tolerable. It was on the whole quite encouraging." Turbulent ablation remained one of the unknowns that could only be decided by a nuclear test. "We just said, 'We'll see when we do the trials whether that happens or not.'"[53]

THE BOLO AND THE SQUID

"Everybody did a little of everything," recalled Dyson about the project's initial year. "There was no division of the staff into physicists and engineers. I particularly enjoyed being immersed in the ethos of engineering, which is very different from that of physics. A good physicist is a person with original ideas. A good engineer is a person who makes a design that works with as few original ideas as possible."[54] The collective nature of the project offered a welcome escape from the personal competition that dominated theoretical physics, with the group operating more in the spirit of a team of mountaineers.

Dyson personally wrote at least eighteen technical reports, as well as much of the first year's still-classified 197-page progress report.[55] *Supersonic Flow Past an Edge* addressed the question of whether the pusher plate might suffer catastrophic instabilities at its perimeter, while *Stability and Control of Space Vehicle* concerned how to maintain stability when misalignment of an explosion produces an eccentric thrust.[56] *Optimal Programming for Vertical Ascent of a Space-ship in the Atmosphere* concluded that "the starting velocity $v(0)$ at height zero is not zero," a requirement that became manifest when the group built a one-meter-diameter, high-explosive-propelled flying model, which was given its initial acceleration by a one-pound charge of black gunpowder that lifted it out of a shallow tub, before subsequent two-pound charges of C-4 high explosive were ejected by the ascending ship.[57] The delivery and ejection of the bombs at precise intervals without malfunction was so critical that engineers from Coca-Cola bottling plants were brought in to consult.

The shock absorber problem remained unresolved. The 1,000 ton pusher plate would be given a 1,000 g acceleration by each explosion, reduced by a factor of about ten by the primary, pneumatic shock absorbers, transmitting a 100 g acceleration to the intermediate platform, where the secondary, hydraulic shock absorbers would reduce the acceleration by another order of magnitude and deliver a tolerable impulse to the inhabited section of the ship. The vehicle constituted a three-mass,

Freeman Dyson (right, holding briefcase) watching preparations for a tethered test of a one-meter-diameter flying model propelled by high explosives at General Atomic's Point Loma test site, summer 1959. Source: Photo by Jaromir Astl, courtesy George Dyson.

two-spring dynamic system, requiring precise control of pulse yield and timing to be kept in phase.

In a still-classified report, *The Bolo and the Squid*, Dyson suggested a different approach.[58] "It's a way of getting rid of the shock absorbers," he explained:

The shock absorbers have always been the most difficult part of the design. So what you would like to do is have a pusher plate which is only attached by a string to the rest of the ship. You bang the pusher plate and it starts moving at a high speed but you don't try to absorb the shock in a mechanical structure, you just absorb it in a long cable that connects the pusher plate with the rest of the ship. The Bolo is just two pusher plates with a long string between them. You bang on the right hand one and it swings around [and] then you bang on the left one and it swings back. You are getting all this momentum each time and if you have a cable that is a kilometer long, it's like having a shock absorber with a stroke of a kilometer so in principle it's very, very nice. The ship is anywhere where it doesn't get the full blast, halfway along the cable is probably a good place. The squid was just a more sophisticated version of that in which you had a pusher plate attached by four cables, on the order of a mile long, and the whole thing is spinning around so the cables are pulled out just by centrifugal force, and the ship is sitting behind, far enough away so it doesn't get much of the blast.[59]

The Bolo and the Squid was followed by Dyson's *Dimensional Study of Orion-Type Spaceships*, taking a wide-ranging look at the sizes of vehicles that could take off from the ground. "The dimensional study was less serious," Dyson explained, "but it answers the question, 'Did you explore the outer limits of this technology?' The answer is yes."[60] He concluded that "ships able to take off from the ground and escape from the Earth's gravitational field are feasible with total masses ranging from a few hundred to a few million tons. The payloads also range from zero to a few million tons. The number of bombs to be carried is independent of the size. The total cost of each trip in fissionable material and in atmospheric contamination is also roughly independent of the size."[61]

Dyson considered *Radioactive Fall-Out from Bomb-Propelled Spaceships*, yet to be declassified, to be his most important report.[62] "I remember working on the fall-out problem as my main responsibility for some months," he noted. "I had become convinced by the summer of 1959 that the long-term acceptability of Orion would depend on the availability of pure or almost-pure fusion bombs. That was why I went to Livermore that summer for two weeks, to find out whether such bombs were possible or probable. At that time the prospects for pure fusion bombs looked promising. Fortunately, as it turned out later, none of the ideas that Livermore was considering worked. All the practical designs for Orion were based on ordinary fission bombs. And for me that was a fatal flaw."[63]

"How many people would die of cancer from the Orion fallout?," Dyson had asked himself. "That was for me the most important question to be answered, and I tried hard to answer it honestly. Unfortunately I do not know whether my estimates of numbers of deaths were ever included in an official report. Most likely, the managers of the project made sure that my numbers never appeared in documents that outside critics might read, for the same reason that discussion of crew survival rates never appeared in any documents that we wrote at RAF [Royal Air Force] Bomber Command during World War II."[64]

"SATURN BY 1970"

The initial plan was to launch a four-thousand-ton ship on a shake-down cruise to Mars in 1965, taking off either from Jackass Flats at the Nevada test site or from a barge in the Pacific that could be washed down or sunk after the launch. The first two hundred explosions, fired at half-second intervals, would lift the ship from sea level to 125,000 feet. Each kick adds about twenty miles per hour to the ship's velocity, an impulse equivalent to dropping the ship from a height of fifteen feet. Six hundred more explosions, gradually increasing in yield, would loft the ship into a three-hundred-mile orbit around Earth. "I used to have a lot of dreams about watching the vertical flight," remembered Taylor. "The first flight of that thing doing its full mission would be the most spectacular thing that humans had ever seen."[65]

Scientists and engineers from fourteen different countries joined the project, led by Taylor's conviction that national boundaries would dissolve in space. A crew of between fifty and a hundred, divided evenly between Air Force officers and civilian scientists, would land on Mars for a year of surface exploration before returning to low Earth orbit where the ship could be refurbished or retired as an orbital base. Much of the mass that would be expended on the voyage was inert propellant that served, once Earth's atmosphere was left behind, to translate the energy released by the nuclear explosions into the momentum required to drive the ship. If a destination was chosen where propellant could be gathered for the return trip, a more ambitious mission could be attempted with the same takeoff mass.

The moons of Saturn promised both suitable propellant and a velocity match with the arriving ship, allowing a landing with minimal deceleration and without the gravity losses required to land on Mars. "We knew very little about the satellites in those days," recalled Dyson. "Enceladus looked particularly good. It was known to have a density of .618, so it clearly had to be made of ice plus hydrocarbons, really light things which were what you need both for biology and for propellant."[66] In August 1958, General Atomic issued Dyson's twelve-page classified report, *Trips to Satellites of the Outer Planets*, outlining the size of ship, velocity increments, nuclear yields, and propellant budget that would be required for the trip.

The first-generation, twenty-kilometer-per-second ship designed to land on Mars could become a forty-kilometer-per-second ship that would land on Enceladus, replenish its expended propellant, and return to Earth. "Saturn by 1970" became the motto of the group. "The Orion system, peculiarly well suited to take maximum advantage of the laws of celestial mechanics, makes possible round trips to satellites of Jupiter in two years or to satellites of Saturn in three years, with takeoff and landing on the ground at both ends," Dyson concluded. "Using the outer planets as hitching-posts, we can make round trips to their satellites with overall velocity increments which are spectacularly small. The probability that we can refuel with propellant on the satellites makes such trips hardly more formidable than voyages to Mars."[67]

Dyson argued for a voyage "like Darwin's *Beagle*, which took five years."[68] Taylor expected a mixed crew, including children, on the voyage and came up with a design for a chamber to enable weightless sex. "What can you do that you can't do on the ground?" he asked.[69] "The mission was the grand tour of the solar system," remembered Harris Mayer, who, unlike Taylor and Dyson, did not plan to go along. "I knew enough about the space business, and this was dangerous," he said. "You had to be crazy to go." Did Dyson intend to go? "Oh, yes, he wanted to go," said Mayer. "Enceladus was the one he wanted to go to then. What was remarkable is that he would come and talk to me and it was as if everything were all done."[70]

DEEP SPACE FORCE

On 31 October 1958, after a flurry of last-minute explosions on both sides, the United States and the Soviet Union entered into a moratorium on nuclear testing that lasted until 1 September 1961, when the Soviets broke the cease-fire with the first of forty-five explosions in sixty-five days. Fourteen of these yielded above a megaton, with one yielding sixty-three megatons, an explosion so large that it blew a hole in the atmosphere, radiating much of its energy into space. The moratorium had worked in Orion's favor: with testing on hold, talent at the weapons labs was idled and General Atomic could pick up the slack. To some, Project Orion was an interplanetary spaceship; to others, it was the ultimate in weapons-effects tests: keeping some of the best physicists working on improved bomb designs and computer codes while the moratorium was in effect. "We should think of Orion as paving its own way through the nuclear explosion test ban, rather than simply waiting for it to disappear," Taylor noted at the time.[71]

The Air Force Special Weapons Center (AFSWC) in Albuquerque kept Orion on life support, shielding the project from those who sided with NASA in trying to shut it down. "Orion was a way to put a burr under the air force saddle blanket," explained Brian Dunne.[72] "When you would go out privately with people in the air force, and talk about what's Orion for," added Taylor, "it was to explore space, no question about that."[73] To fend off NASA and justify allocation of Air Force funds, AFSWC required a military mission. They named it Deep Space Force.

By 1960, the global nuclear stockpile had reached 30 million kilotons: ten thousand times the total firepower expended in the Second World War. Deep Space Force would have stationed a fleet of heavily armed Orion ships in orbits beyond the moon, allowing more time to consider or reconsider a retaliatory strike. "On the order of 20 space ships would be deployed on a long-term basis," a declassified summary explains:

At these altitudes, an enemy attack would require a day or more from launch to engagement. Assuming an enemy would find it necessary to attempt destruction of this force simultaneously with an attack on planetary targets, initiation of an attack against the deep space force would provide the United States with a relatively long early warning of an impending attack against its planetary forces. Because of its remote station, the force would require on the order of 10 hours to carry out a strike, thereby providing a valid argument that such a force is useful as a retaliatory force only. This also provides insurance against an accidental attack which could not be recalled.

Crews of twenty to thirty, enjoying "an Earth-like shirt sleeve environment with artificial gravity," would be "deployed alternately, similar to the Blue and Gold team concept used for the Polaris submarines" in orbits about 250,000 miles from Earth.[74]

Dyson welcomed the support of the Air Force officers, led by Captain Donald M. Mixson, who were risking their careers to keep Orion afloat, but he believed their strategy suffered from some of the same misconceptions that had led to such heavy losses among the Lancaster squadrons at Bomber Command. "The bewitching picture of Manned Space Forces sailing the seas of space in silent and secret majesty, like the grey ships of the British fleet which Napoleon never saw but which defeated his dreams of empire, is an illusion," he argued in 1962. "Space is an environment in which it is extremely difficult to hide."[75] Mixson countered that Deep Space Force had been developed "not to make Orion a military machine, but to con a military machine into yet another installment of funds to keep your beautiful big dream alive. You see, I shared that same dream and it was the only reason I was in the Air Force [when] NASA did not [yet] exist."[76]

Orion became a pawn in a rivalry that went back centuries: whether wars should be decided by admirals commanding ships at sea or by generals commanding armies on the ground. President Kennedy bristled at General Thomas Power's statement, as commander of the Strategic Air Command, that "whoever controls Orion will control the world." The army, which had enlisted Wernher von Braun and his chemical rockets, won: the moon was in, but Mars and Saturn were out. "Don't try to be kind to [U.S. Secretary of Defense Robert] McNamara," Herbert York advised when Dyson asked him about assigning blame for Orion's demise.[77]

DEATH OF A PROJECT

After the Limited Nuclear Test Ban Treaty was signed in 1963, Taylor kept up the hope of launching an interplanetary mission as a treaty-compliant joint venture with the Soviet Union. "The effect of the treaty on Orion is likely to be quite healthy in the long run," Dyson wrote to Taylor. "It means the emphasis is automatically pushed away from military applications and toward long range exploring missions. It is just what we always wanted to do."[78] Even Niels Bohr lent his support. But without support from NASA, this proposal stalled.

American politics, not international treaties, brought Orion to a halt. The project had been orphaned almost from the start. The military was unable to adopt a project aimed at peacefully exploring the solar system while NASA was unwilling to adopt a project driven by bombs. Dyson and Taylor envisioned a NASA-USAF collaboration, the way James Cook had explored the Pacific in Royal Navy vessels, or the US Navy provides Antarctic logistics while the National Science Foundation supplies the scientific mission and personnel. They underestimated the bitter rivalries over space. It was NASA's refusal to endorse an interplanetary mission in ships that would belong to the Air Force that kept the project on the ground. According to NASA deputy Earl Hilburn, the reason for NASA's final vote to kill the project was that "established policy is that NASA has no requirement for manned planetary missions . . . and if a big expansion is wanted it must be a political decision of Congress, not an internal NASA question."[79]

Orion never gained the allies it needed in Congress, partly because of secrecy and partly because General Atomic could not compete against the NASA tactic of dividing its projects up across as many congressional districts as possible to ensure support. David Weiss, a former test pilot brought in to start looking at getting a flight test program off the ground, explained that de Hoffmann "wanted to build the whole thing himself, like in a nineteenth-century shipyard, where you had your own foundry to make castings for all the metal parts."[80]

By February 1965 the Orion group at General Atomic was down to nine and on 30 June 1965, Major John O. Berga of the Air Force Weapons Laboratory issued a terse, one-page plan change, concluding "This Project is hereby terminated."[81] Orion, caught between the hopes of the 1950s and the fears of the 1960s, was at an end. The peaceful mission was abandoned, yet military programs that originated with Project Orion—directed-energy nuclear weapons, electromagnetic-pulse weapons and countermeasures, and antiablation defenses against antisatellite weapons—have continued to this day.

Dyson left the project in fall 1959 once it became clear that NASA would not be signing on. He returned to La Jolla for a sabbatical year in 1964–1965, but had nothing further to do with the project except to write an obituary, published in *Science* under the title "Death of a Project," with a one-sentence abstract: "Research is stopped on a system of space propulsion which broke all the rules of the political game."[82] Only now could he write publicly about a project that should have been in the open all along. Orion's opponents resented the attention and made sure that any chance of restoring funding was withdrawn. One of these critics "clearly wishes ORION not only to stay dead, but quiet also," a memorandum from the project's advocates at AFSWC complained.[83]

"The men who began the project in 1958," Dyson reported, "aimed to create a propulsion system commensurate with the real size of the task of exploring the solar system, at a cost which would be politically acceptable, and they believe they have demonstrated the way to do it."[84] But once NASA chose the path of chemical rockets, "there was no more brave talk of manned expeditions to Mars by 1965, and of sampling the rings of Saturn by 1970. What would have happened to us if the government had given full support to us in 1959, as it did to a similar bunch of amateurs in Los Alamos in 1943? Would we have achieved by now a cheap and rapid transportation system extending all over the Solar System? Or are we lucky to have our dreams intact?"[85]

PANDORA'S BOX

In 2012, Dyson revisited the questions of "whether I believe Orion has a future . . . whether I share a hope that some new version of Orion might take us to the stars . . . [and] whether our dreams of fifty years ago are dead. The answer to all three questions is no."[86] There are now better ways to explore the solar system, and even Orion is too slow to reach the stars.

Dyson's preferred alternative had become laser propulsion, with the energy source remaining safely on the ground while a lightweight and unencumbered vehicle, using an inert material like water for propellant, rides a pulsed laser beam into orbit or beyond. The reaction as the propellant is vaporized by the laser beam and delivers an impulse to the vehicle is reminiscent of Orion, as Dyson noted in a 1977 study, dusting off old skills in writing the report. "In the [proposed] design the hot propellant gas is in contact with the vehicle for a much smaller fraction of the duty cycle," he wrote. "For example, if the contact lasts 100 microseconds for each pulse and the pulse rate is 100 per second, the contact occupies only 1% of the cycle. The exposed surface is mainly a flat plate."[87]

NASA still stood in the way. "The basic ground-rule of the NASA study contracts is that laser propulsion must be used as an adjunct to the Space Shuttle and not as a replacement for the Shuttle. This means that laser propulsion must be used to move from a low earth orbit (LEO) to a high earth orbit (HEO) but must not be used for launch from the ground," Dyson complained. "It may well turn out that it is cheaper to launch a payload from the ground directly into HEO than to move the same payload from LEO to HEO. This is a possibility which the NASA studies refuse to contemplate."[88]

After Project Orion came to an end, Taylor devoted himself to disarmament, the restoration of Earth through ice ponds and solar energy, and to raising the alarm about home-grown nuclear bombs. "The use of small numbers of covertly-delivered nuclear explosives by groups of people that are not clearly identified with a national government is more probable, in the near future, than the open use of nuclear weapons by a nation for military purposes," he warned in 1966. "The group could be an extremist group of US citizens who believe they are trying to save the U.S."[89]

Taylor believed we had only explored a small fraction of possible nuclear weapon designs, and this kept him awake at night. "I had a dream last night, about a new form of nuclear weapon, and I'm not telling anybody what this is, because I'm really scared of it," he revealed in a conversation a few years before his death. "I have tried, I thought successfully, to hold on to a vow of just not thinking about new types of nuclear weapons any more. And what's happened is that it has gone from my conscious to my unconscious, and it's emerging as a dream; I cannot shut it off."

"We haven't opened Pandora's box and found the really important content of that box which was hope," Taylor continued. "Down in the bottom. If we just fiddle around at the surface, all kinds of terrible things come out."[90] For Taylor and Dyson, there was a brief moment in history when Project Orion offered that hope. Both decided it was time to close the box.

"I also have a bit of Ted in me," Dyson revealed shortly after Taylor's death. "Recently I could not stop myself from thinking about a new kind of nuclear device that could really make things worse for everybody. I am determined to carry this with me to the grave."[91]

NOTES

1. Freeman Dyson, interview, 1 August 1993.

2. Dyson, interview, 1 August 1993.

3. Freeman Dyson, "Mankind in the universe," lecture given at the meeting of the German and Austrian Physical Societies, Salzburg, Austria, 29 September 1969.

4. Dyson, interview, 1 August 1993.

5. Freeman Dyson, *Disturbing the Universe* (New York: Harper & Row, 1979), 53.

6. Frederic de Hoffmann, "A novel apprenticeship," in *All in Our Time*, ed. Jane Wilson (Chicago: Bulletin of the Atomic Scientists, 1974), 171.

7. Edward Teller, interview, 22 April 1999.

8. Brian Dunne, interview, 11 February 1999.

9. Dunne, interview, 13 July 1998.

10. Dyson to parents, 12 August 1956.

11. Freeman J. Dyson, *Two-Group Treatment of the Warm Neutron Effect*, GAMD-6, 17 September 1956. [GAMD = General Atomic Manuscript Document]

12. Dunne, interview, 11 February 1999.

13. Dunne, interview, 11 February 1999. The fission reactions within the TRIGA reactor yielded high-energy electrons among the decay products. The reactor was immersed in water to help keep it cool; within the water, the emitted electrons traveled faster than light (which travels more slowly in water than in a vacuum or ordinary air). Under such conditions, light emitted from the moving electrons (known as Cherenkov radiation) yields a blue glow.

14. Theodore Brewster Taylor, Andrew Wetherbee McReynolds, and Freeman John Dyson, "Reactor with Prompt Negative Temperature Coefficient and Fuel Element Therefor," U. S. Patent No. 3,127,325, filed 9 May 1958, issued 31 March 1964.

15. Freeman Dyson to Jeffrey Bezos, 22 November 2016.

16. Dyson, interview, 11 May 1998.

17. Dyson, interview, 11 May 1998.

18. Freeman J. Dyson, "Experiments with Bomb-Propelled Spaceship Models," in *Adventures in Experimental Physics*, ed. Bogdan Maglich (Princeton, NJ: World Science Education, 1972), 324.

19. Theodore B. Taylor, interview, 11 August 2000.

20. Taylor, interview, 11 May 1998.

21. Taylor, interview, 2 July 1999.

22. C. J. Everett and S. M. Ulam, "Nuclear propelled vehicle, such as a rocket," British Patent Specification No. 877,392, application filed 17 February 1960, issued 13 September 1961. Application filed in the United States on 3 March 1959.

23. C. J. Everett and S. M. Ulam, *On a Method of Propulsion of Projectiles by Means of External Nuclear Explosions*, Los Alamos Scientific Laboratory Report LAMS-1955 (August 1955), 3–5.

24. Taylor, interview, 9 May 1998.

25. Dyson, interview, 11 May 1998.

26. Dyson to parents, 1 January 1958.

27. Dyson to parents, 11 January 1958.

28. Major Lew Allen, USAF Office of Special Projects, to Ray DeGraff, Air Research and Development Command, 29 May 1958, in "Project Orion: An Air Force Bid for Role in Aerospace," 161–297 of the *1964 Annual History of the Air Force Weapons Laboratory, 1 January—December 1964*, compiled by Dr. Ward Alan Minge, Captain Harrell Roberts, and Sergeant Thomas L. Suminski, 165, AFSWC.

29. Lew Allen, interview, 10 June 1999.

30. "Project Orion: An Air Force Bid for Role in Aerospace," 169.

31. Herbert F. York, interview, 6 February 1999.

32. York, interview, 6 February 1999.

33. York, interview, 6 February 1999.

34. Air Force Contract AF 18(600)-1812, "Feasibility Study of a Nuclear Bomb Propelled Space Vehicle," 30 June 1958, Exhibit "A"—Statement of Work, 1.

35. Don Prickett and Ed Giller, interview, 16 September 1999.

36. Taylor, interview, 10 November 1999.

37. General Atomic, news release, Washington, DC, 2 July 1958.

38. Dyson to J. Robert Oppenheimer, 4 July 1958.

39. Dyson to parents, 1 July 1958.

40. Taylor, interview, 8 May 1998.

41. Freeman J. Dyson, *Free Expansion of a Gas, Part 2, Gaussian Model.* GAMD-507, 5 September 1958.

42. Dyson, interview, 12 July 1998.

43. Dyson, interview, 11 May 1998.

44. Dyson, interview, 11 May 1998.

45. Taylor, interview 7 November 1999.

46. Dyson, interview, 16 March 2000.

47. Dyson, interview, 11 May 1998.

48. Dyson, interview, 11 May 1998.

49. Dyson, *A Note on Maximum Opacity*, GAMD-469, 8 July 1958.

50. Dyson, interview, 11 May 1998.

51. Dyson, interview, 11 May 1998.

52. Dyson, *Preliminary Study of Convective Ablation*, GAMD-710, 17 March 1959.

53. Dyson, interview, 11 May 1998.

54. Dyson, "Experiments with Bomb-Propelled Spaceship Models," in *Adventures in Experimental Physics*, ed. Bogdan Maglich (Princeton, NJ: World Science Education, 1972), 326.

55. Brian B. Dunne, Freeman J. Dyson, Michael Treshow, and Theodore B. Taylor, *Project Orion: Feasibility Study of a Nuclear-Bomb-Propelled Space Vehicle, Interim Annual Report, 1 July 1958–1 June 1959*, GAMD-837, April 1959.

56. Freeman J. Dyson, *Supersonic Flow Past an Edge*, GAMD-565, 6 October 1958; Freeman J. Dyson, *Stability and Control of Space Vehicle*, GAMD-424, 8 July 1958.

57. Freeman J. Dyson, *Optimal Programming for Vertical Ascent of a Space-ship in the Atmosphere*, GAMD-619, 11 December 1958, 2.

58. Freeman J. Dyson, *The Bolo and the Squid*, GAMD-599, 14 November 1958.

59. Dyson, interview, 12 July 1998.

60. Dyson, email to the author, 21 June 1999.

61. Freeman J. Dyson, *Dimensional Study of Orion-Type Spaceships*, GAMD-784, 23 April 1959, 2–3.

62. Freeman J. Dyson, *Radioactive Fall-Out from Bomb-Propelled Spaceships*, GAMD-835, 2 June 1959.

63. Freeman Dyson, email to the author, 27 December 1998.

64. Dyson, email to the author, 15 December 1999.

65. Theodore B. Taylor, interview, 8 November 1999.

66. Dyson, interview, 12 July 1998.

67. Freeman J. Dyson, *Trips to Satellites of the Outer Planets*, GAMD-490, 20 August 1958, 12.

68. Dyson, interview, 12 July 1998.

69. Taylor, interview, 8 May 1998.

70. Harris Mayer, interview, 17 September 1999.

71. Theodore B. Taylor, unpublished journal, 3 October 1960.

72. Dunne, interview, 10 November 1999.

73. Taylor, interview, 9 November 1999.

74. General Atomic, *Potential Military Applications*, GA-C-962, 1 March 1965, 16–21.

75. Freeman J. Dyson, *Implications of New Weapons Systems for Strategic Policy and Disarmament: Program for a Study*. Arms Control and Disarmament Agency, Washington, DC, 14 August 1962. [Unpublished/Secret]

76. Donald M. Mixson to Freeman J. Dyson, 23 October 1979.

77. Freeman Dyson, notes from telephone conversation with Herbert York, appended to letter to Herbert York, 3 March 1965.

78. Dyson to Theodore B. Taylor, 11 November 1963.

79. Freeman J. Dyson, "ORION: Notes on History," undated notes from conversation with James Nance, early 1965.

80. David Weiss, interview, 12 June 1999.

81. John O. Berga, Major, USAF, *Research and Technology Resume, Nuclear Impulse Propulsion Technology Studies, Plan Change*, 30 June 1965.

82. Freeman J. Dyson, "Death of a Project," *Science* 149, no. 3680 (9 July 1965): 141.

83. Captain Ronald F. Prater, USAF Project Engineer, *Memorandum for the Record: Sponsorship of ORION Development Planning*, 22 July 1965.

84. Dyson, "Death of a Project," 141.

85. Freeman J. Dyson, "Experiments with Bomb-Propelled Spaceship Models," in *Adventures in Experimental Physics*, ed. Bogdan Maglich (Princeton, NJ: World Science Education, 1972), 326.

86. Freeman Dyson, unpublished draft of a new foreword to *Project Orion*, 16 September 2012.

87. Freeman J. Dyson and F. W. Perkins Jr., *Jason Laser Propulsion Study, Summer 1977*, SRI Technical Report JSR-77–1, December 1977, 8.

88. Dyson and Perkins, *Jason Laser Propulsion Study*, 17–18.

89. Theodore B. Taylor, "Notes on criminal or terrorist uses of nuclear explosives," unpublished typescript, 7 November 1966.

90. Taylor, interview, 10 December 1999.

91. Dyson, personal communication, 4 May 2006.

6 DYSON, WARFARE, AND THE JASONS

ANN FINKBEINER

We are scientists second, and human beings first. We become politically involved because knowledge implies responsibility.
—Freeman Dyson, *Disturbing the Universe* (1976)

Somewhere around 1960, Freeman Dyson was invited to join a new group of scientists just setting up as an independent government contractor. The group called itself Jason; it proposed to answer the Department of Defense's technical and scientific questions about, for example, ballistic missile defense, antisubmarine warfare, and the detection and consequences of nuclear explosions.[1] Jason is still, at the time of publication sixty years later, in business; Dyson joined and remained a member until his death. In the career of a mathematically inclined quantum physicist, such dedication to the Defense Department's technical problems might seem like a hard-left turn. But if it was incongruous, it was no surprise. Alongside his academic career, Dyson had another one in the interwoven national security issues of disarmament, nuclear test bans, and arms control.

Dyson was primed for this second career. He had grown up in an atmosphere saturated with the First World War: his father served in the

British army; so did his uncle Freeman, who was his mother's brother, his father's friend, and his own namesake, whose death in the war at age thirty-three was "the central tragic event in our family history."[2] So as a young man, based on what he knew of the carnage of the First World War, he watched the approach of the Second World War expecting that he and all his friends would be killed. At age fifteen, he counted the names on plaques commemorating his school's six hundred war dead, and calculated that given graduating classes of eighty each, seven years' worth of students had been wiped out. "We of the class of 1941 were no fools," he wrote later. "We knew how to figure the odds." They gave themselves one-in-ten odds that in the next five years, they'd be dead.[3]

As a result, for a while he and his friends became fierce pacifists, but Dyson quickly realized that the example of the French Vichy government meant that the English people, if invaded, would also have to either collaborate with the enemy or resist it.[4] He chose resistance. He joined the army's Officers Training Corps and at age nineteen was assigned to the Bomber Command, where he learned from experience that the military often didn't listen to its scientists and often made lethally stupid decisions. "Bomber Command gave me a lifelong commitment to making sure this sort of tragedy doesn't happen again," he later told an interviewer. "We need a military, but we should have a military that is sane and does sensible things and talks to the outside world."[5]

His writing about this part of his early life is as clear and reasonable as all of his writing but it's unusually intense, as though he'd taken war and its consequences to heart and always kept them there. So two years later, in 1947, at Cornell University, he was ready to meet his professors and colleagues—like Hans Bethe and Philip Morrison—who were graduates of the Manhattan Project and still reckoning with their own roles in creating nuclear weapons and consequently also had reasons for wanting to talk to the military.

Bethe had just helped found the Federation of Atomic [later, American] Scientists (FAS), and Dyson immediately began going to meetings, enjoying "all these energetic people getting involved with political campaigns."[6] FAS was then at the center of scientists' political attempts, "passionately" and "persistently," to ensure that control of nuclear weapons would not be military and national but civilian and international.[7] Their efforts had failed to ensure international control but had succeeded with civilian: with the passage of the McMahon bill in 1946, control of nuclear weapons was given not to the Defense Department but to a new, civilian organization, the Atomic Energy Commission.[8]

Later, in 1960, Dyson officially joined FAS, where he proposed that it should advocate for a policy of no first use of nuclear weapons. The FAS agreed and, to Dyson's gratification, backed no first use with enthusiasm; Dyson was also astonished to learn "how seriously the people high in the government take our opinions."[9] In 1962, he was bemusedly elected to be its chairman and served for the year allotted.[10] "I have not been an outstanding chairman, but my time in office has gone by without FAS falling apart," he wrote to his family. "My talent does not lie in the business of organizing meetings and committees. It was worthwhile to try it once, but it is not my metier."[11]

After leaving Cornell but before joining FAS, Dyson had become fairly expert in nuclear weapons himself, embarking on two projects in part because he (along with much of the physics community) wanted to find something to do with nuclear energy besides kill people. His appointment at the Institute for Advanced Study meant that his summers were free, and he used them, beginning in 1956, to work at General Atomic in La Jolla, California, on a commercial nuclear reactor that produced radioactive isotopes for nonspecialists to use in hospitals, industry, and research. Dyson designed the reactor so that if it began a chain reaction leading to an explosion, it would, by the laws of physics, shut itself down automatically.[12] The reactor, called TRIGA (for Training, Research, Isotopes, General Atomics), has been solidly successful: sixty-six TRIGAs have been built in twenty-four countries and it's the most widely used research reactor in the world.[13]

The second project, begun in 1958 also at General Atomic, was Orion, a spaceship powered by small nuclear explosions.[14] Though a peaceful use of nuclear weapons, Orion meant Dyson had to engage in a fancy logical finesse over nuclear test bans that took years to untangle. He had originally proposed the no first use policy to the FAS as an alternative to a policy they considered backing, of banning nuclear tests.[15] But Orion depended on nuclear explosions in the atmosphere, and without them, he wrote later, "Orion could not survive."[16] So in 1960, Dyson published an article, "The Future Development of Nuclear Weapons," in the influential journal *Foreign Affairs*.[17] He argued that a test ban would be a "dangerous illusion" because the future was in weapons not yet developed, particularly low-fallout bombs called "fission free" (fusion bombs that would not require fission triggers) that he referred to as the "wave of the future," and a country that had signed a treaty to ban all nuclear tests "is likely to find itself in the position of the Polish army in 1939, fighting tanks with horses."[18] The *Foreign Affairs* article was covered in the press. "I had some busy days," he told his family, "with reporters calling from Washington and photographers coming to the Institute to take my picture."[19]

He knew about low-fallout, fission-free bombs because in 1958, he had spent a "wildly exciting" week working on them at Lawrence Livermore National Laboratory in California, trying to design a bomb that would lower the fallout from an Orion launch.[20] He was also incidentally helping colleagues at Livermore develop a neutron bomb that would kill people but leave infrastructure intact.[21] His logical finesse in support of Orion began crumbling: "To that extent I share the responsibility for the existence of neutron bombs" he wrote later, he could never thereafter claim that "the bombs we wanted to use for Orion had nothing to do with bombs that are designed for killing people."[22]

His finesse crumbled completely after he spent another two summers, in 1962 and 1963, at the newly created Arms Control and Disarmament Agency (ACDA), an independent US government agency—"a delightful place to work," he said, too small and new to have yet developed a bureaucracy.[23] ACDA was in the process of preparing for international negotiations on a nuclear test ban and nuclear disarmament. Because

Dyson had published his opposition to a test ban, he was asked to work instead on disarmament; during one summer, he came up with a plan for nations to disarm by geographical zones. He also studied Russian news and government publications (he had learned to read Russian), concluding that Russians negotiated disarmament by bluff; his memoranda on both were apparently dismissed.[24]

During the next summer, however, a test ban was imminent and ACDA needed all hands on deck, so Dyson worked on the test ban. He had a tiny "moment of glory." A final problem in international negotiations was whether the ban should include peaceful atmospheric explosions; the US negotiators called the ACDA for advice on whether the country should give up its objection to the inclusion. The ACDA office was all but empty, and it was Dyson, who knew the reason for the US objection was local politics, who said, "Of course we should give way."[25] In 1963, when the Limited Test Ban Treaty came before the US Senate Foreign Relations Committee, Dyson testified in favor and the ban was ratified.

What changed? Why was Dyson now in favor of a test ban? In part, Orion was all but dead by then, and in part, during the first summer at ACDA, Dyson had plotted the number of nuclear tests by all countries from 1945 to 1962 and found that the number of tests each year was rising exponentially, doubling every three years.[26] So the choice was not between banning and not banning tests; it was between banning tests and watching them rise exponentially "ad infinitum."[27] "At some point," Dyson thought, "we have to stop." He weighed a test ban against nuclear-powered space travel, and the test ban won. In 1962 he wrote a memorandum to the director of ACDA explaining the exponential rise and his conclusions. Later, in 1979, he wrote that in retrospect, his *Foreign Affairs* article had been a mistake, a "political diatribe," "a desperate attempt to salvage an untenable position with spurious emotional claptrap."[28] (His colleagues in academia never forgot it, however, and the anti-test-ban label stuck to him for decades.)

After 1963, ACDA had its own staff and no longer needed freelancers like Dyson, leaving his summers free for the newly formed group

Year	No of Tests	Cumulative Total Number of Tests	Comparison Series
1945	3	3	7
1946	2	5	9
1947	0	5	12
1948	3	8	15
1949	1	9	19
1950	0	9	24
1951	18	27	30
1952	11	38	38
1953	15	53	48
1954	7	60	60
1955	19	79	76
1956	26	105	96
1957	44	149	121
1958	82	231	152
1959	0	231	192
1960	3	234	242
1961	60	294 274	304
1962	63	357 (400) 403	384

July 26 1963 NYT Figures

(414)

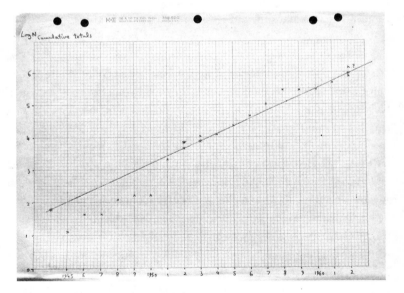

Freeman Dyson's table and hand-drawn logarithmic plot from July 1962, showing the cumulative number of nuclear weapons tests to date. The exponential growth of such tests over time convinced Dyson to support a test ban. Source: Freeman Dyson, memorandum to Frank Long, US Arms Control and Disarmament Agency, 31 July 1962, courtesy George Dyson.

of scientists with top-secret clearances called Jason. Thereafter Jason became the primary venue for Dyson's interests in national security.

A NEW JASON

Jason might have been designed with Dyson in mind. It suited his tastes for variety and the mathematically ascertainable, his concerns with the military, and his unorthodox and contrarian character. The group met during summers; the studies were solely about technology or science, not policy. At the time, the group did studies mainly for the Defense Department. "I always have felt it was good, from the Bomber Command experience," Dyson told an interviewer, "to talk to the soldiers."[29] Jason worked for other science-related departments and agencies as well, so its studies were widely varied. The group's organization was loose, and though it tightened with time, it has never seemed to be more organized than necessary; but it did have committees with responsibility for, say, making administrative decisions or creating liaisons with sponsors. Dyson said he never served on any of them: "The fact is, when I'm on a committee, I usually go to sleep," he said. "Some people are just no good at committees."[30]

Most important, Jason was, and is, utterly independent. Its membership has always been dominated by academic scientists with tenure and no conflicting interests. The group depends on no single sponsor; the Jasons say, "We're not beholden." The sponsor and Jason work out the subjects for the studies together, but ultimately, Jason chooses which questions to undertake and does not guarantee that the sponsors will like the answers.[31]

Dyson's letters home and the sparse documentation from Jason show Dyson agreeing to join in 1960, meaning he was one of the first to be invited, but he himself thought he deferred being an active member until after his term at ACDA was over, in 1965.[32] Dyson was invited by Marvin (Murph) Goldberger, then a physicist at Princeton and one of Jason's founders, whom Dyson had met at General Atomic. Dyson said he joined for typically double-edged reasons: during the summers,

Princeton was a ghost town, as was the Institute; he already had clearance for top-secret work; he had four children by then and the contract money was welcome; and, he said, doing studies with Jason was better than playing golf. And now that TRIGA and Orion were in the past, he needed practical problems to work on. "I'm basically a tinkerer," he said. "I just, I like to solve problems."[33] These reasons could not have been the whole story: he attended the Jason studies, eventually held in La Jolla, for most, if not all, of the seven-week summer sessions for the next fifty-plus years. "I have stayed with JASON all my life ever since," he told an interviewer, "it's been a big part of my life."[34]

The members of Jason when Dyson joined were mostly physicists, and at the time, their studies were basically categories of physics problems: the science fundamental to test ban treaties and necessary to antisubmarine

Marvin (Murph) Goldberger and Freeman Dyson at the Institute for Advanced Study in 1979. During the early 1960s Goldberger helped to recruit Dyson to join the Jasons. Source: Photograph by Emmet Gowin and Allen Hess, courtesy Emmet Gowin and the Shelby White and Leon Levy Archives, Institute for Advanced Study, Princeton, New Jersey.

warfare and ballistic missile defense. Could you cheat on a test ban by testing a nuclear bomb underground? What kinds of wavelengths do you need to detect an enemy submarine or communicate with one of your own? Can you detect an incoming enemy missile by the infrared radiation coming from the heat of its launch?[35]

These three general subjects accounted for much of the roughly two hundred studies that Dyson worked on over the years, though his range was wide, and as Jason's range widened with the years, so did his. He didn't come into Jason with any "grand plans," he told an interviewer, "I was generally sort of just hanging around waiting to be told what to do."[36] That probably meant only that he had no specific agenda, not that he didn't choose what to work on. He liked the studies best that were the most scientific, "the ones where you can actually calculate things," he said.[37] He worked on studies of space technology, from hypersonic flight to using ground-based lasers to launch vehicles into space. He worked on counterterror studies, including an assessment of the security of the power grid, and the calculation of the risks of specific categories of terrorism. He liked learning new things, and when, in the 1990s, he began working on biological subjects, such as bioterrorism and mining the data of the Human Genome Project, Dyson joined a Jason-led unofficial laboratory in genetic sequencing (he wasn't very good at it).[38] Beginning in the late 1970s, he turned to Jason's climate studies—work that may have been the origin of his famously contrary views on global warming.[39]

Among the studies he seemed to like the best were what Jason calls "lemon detection." Science in the classified world is done without much peer review from either inside that world or outside it, so classified science risks being, in Dyson's words, "very stupid."[40] Jason often acts as unofficial peer review—it's one reason for its longevity—and its studies have stopped expensive, unscientific defense projects. The best known of these was the Neutrino Detection Primer study. A perennial military suggestion is to detect enemy nuclear submarines by the neutrinos flooding out of their reactors, but every physicist knows that neutrinos rarely interact with ordinary matter and are all but undetectable. Now, says one

Jason, whenever military officials suggest detecting neutrinos again, "we just hand them that report." If a scientific report full of equations can sound exasperated and snippy, this one does; Dyson probably wrote it. He was a little gleeful about the lemon detection studies: they were what Jason was best at, he said, "technically sound arguments for not doing things that were stupid," he told interviewers.[41] "Probably we've saved the government hundreds of billions."[42]

Unlike the Neutrino Detection Primer, most Jason studies are classified, so it's impossible to fully describe them or explain Dyson's contributions to them. In general, he would contribute by, said one Jason member, "thinking about some key tech point in some totally new way that was often pretty specific." His contributions often involved math, "little logico-mathematical gems that he somehow extracted from the muck of a diffuse study topic," said another Jason. Still another said that Dyson was a master at framing mathematical problems: Dyson was as famous among Jasons as among physicists in general for hearing a problem no one could answer and answering it shortly thereafter, leaving the original questioner to figure out unsuccessfully what Dyson had done or what arcane historical solution he had remembered. When the self-selected, self-confident, highly educated Jasons were stuck on a calculation, they took it to Dyson. Set up the calculation right, you can solve it on a laptop or maybe you need giant computers, said one Jason, "but if you have Freeman, you don't need either one."[43]

Jasons work in groups of two to three to ten or more on whatever studies interest them. Dyson often worked alone: he went to most of the briefings by subject matter experts on all the summer's studies; he'd show up at meetings of the groups working on those studies, then say, "Let me think about this part," and go off on his own to solve it. He said he had never been much of a collaborator anyway; a Jason described him as "an inward man."

The late particle physicist Sid Drell told an interviewer that some Jasons could float from one study to the next, but "a guy like Dyson could contribute to all of them at once."[44] Richard Garwin, a Jason member now retired from IBM, said Dyson was often "in deep silence and spoke only when he had something to contribute. Freeman contained infinities."[45]

Given Jason studies' likely classification and the Jasons' reluctance to talk about them, the most concrete way to show Dyson's work with Jason and its intersection with his own interests is to focus on two studies, one on the effects of certain nuclear weapons and the other on a technology for improving the ability to observe things in space. Both were classified at the time but have been opened since.

WEAPONS EFFECTS

Jason had a subgroup of members called the arms controllers—most notably Sid Drell and Richard Garwin—who worked for decades inside and outside Jason, in both the government and public spheres, on the collection of issues going under the name of arms control: test limits or bans, limits on the numbers of nuclear weapons, detection of illegal tests, ballistic missile defense. Though Dyson never again worked in government after ACDA, he did become a public voice on nuclear issues with his book *Weapons and Hope* (1984). In its introduction, he wrote that he was indebted to Drell and Garwin for his education on these issues.[46] As always, Dyson is his own best spokesman and rather than outline what he wrote in *Weapons and Hope*, I'm tempted to advise that you just go read the book because he says it better than I can. I can, however, summarize it briefly, partly for context to Dyson's work in Jason, partly because it was informed by an earlier, odd, and famous Jason study he did.

After the nuclear club of countries built bigger hydrogen bombs, then more and bigger hydrogen bombs until they had massive stockpiles of massive bombs, they avoided global Armageddon by relying on the concept of mutually assured destruction (MAD). Dyson thought MAD was shortsighted.[47] He proposed a concept he called live-and-let-live in which our necessary national security would be guaranteed by conventional and defensive weaponry, and the nuclear weapons in everyone's stockpiles would be used as bargaining chips to reduce all stockpiles to zero.[48]

Dyson's education in nuclear matters came not only from TRIGA, Orion, Drell, and Garwin but also from an ongoing series of Jason studies, beginning in the 1960s, on the signatures and effects of nuclear explosions that were necessary to negotiating treaties to ban nuclear tests. The studies continue today, one or two every year, under the general name of "stockpile stewardship" and are requested by the agency in the Department of Energy that is responsible for the nation's nuclear laboratories (including Los Alamos National Laboratory, Sandia National Laboratory, and Lawrence Livermore National Laboratory), which are themselves responsible for making sure the weapons in the stockpile haven't gotten old and useless but will continue to serve as a deterrent. Jason reviews the individual programs, like the one called the National Ignition Facility that attempts to create thermonuclear fusion with lasers. Stockpile stewardship studies can also provide the evidence necessary for signing test bans. In 1995, a Jason study outlined the science on stockpiled warheads that allowed Secretary of Defense William Perry to sign the Comprehensive Test Ban Treaty to stop testing all nuclear weapons anywhere but still retain confidence in the health of the stockpile.[49] Dyson worked on that 1995 study and on most of Jason's thirty or more stockpile stewardship studies.

Meanwhile, the countries had stockpiled so many massive bombs targeting the enemy's cities and facilities that eventually they ran out of places to target.[50] Rather than leave well enough alone, they had begun adding to their stockpiles smaller bombs that were cheaper, were more easily deliverable, and could be aimed more accurately at military (rather than civilian) targets.[51] These were called tactical nuclear weapons, and the best idea for their use seemed to be that if the enemy used them first, we would counter with ours in what was called a limited nuclear war. Dyson thought this concept too was shortsighted.[52]

He based his opinion on the effects of tactical nuclear weapons, which he had learned about doing a Jason study that no government sponsor had requested. During the Vietnam War, Dyson briefly stopped working on Jason studies. The group was doing several Vietnam-related studies, one of which was on the expensive ineffectiveness of the conventional bombing campaign, and the other, on a plan for an electronic barrier across the Ho Chi Minh Trail that could call in bombers to stop the North Vietnamese from filtering down the trail into South Vietnam and ultimately bring the North Vietnamese to the negotiating table. Dyson had considered joining the barrier project because he preferred defensive strategies to offensive ones, but he decided against it because he thought it wouldn't stop the infiltration (he was right; it didn't). He didn't officially drop out of Jason and didn't think Jason was wrong to continue war work, but he thought the war was stupid and didn't see what he could contribute that would lessen the stupidity.[53]

But sometime in early 1966, Dyson and three other Jason members were at a cocktail party and overheard General Maxwell Taylor, former chair of the Joint Chiefs of Staff and current adviser to President Lyndon Johnson, wonder out loud whether it might be interesting to drop a nuke in Vietnam just to keep the other side guessing.[54] The four Jasons—Dyson, Courtney Wright, Robert Gomer, and Steven Weinberg—knew that the idea of dropping tactical nuclear bombs on the mountain passes used by infiltrators had been floating around the Pentagon.[55] They either knew or learned that the tactical nuclear weapons at the time included bombs that could be sent by artillery or even recoilless rifle.[56] Probably egged on by Gomer, the four decided on their own that they needed to do something; the Jason contract with the government had enough leeway to allow Jasons to work on a problem without being asked.

They ruled out moral or ethical arguments as ineffectual for a military study; as Dyson said, "You can look at this as a purely military question."[57] They read "Oregon Trail," a 1965 multivolume RAND study (still classified) of the wins or losses of historical small-scale wars, plus war game simulations, including limited nuclear wars. First they applied RAND's lessons learned to Vietnam's mountain passes, forests, and supply trails, then to a wider war in Southeast Asia with China and the Soviet Union. They found that local damage—say, blocking a forest supply trail by blowing down all the trees—was considerable but temporary because trees can be blown down only once, and afterward the supply picks up where it left off. More broadly, they found that US military bases made excellent targets and that US interests could be, as their report said, "essentially annihilated."[58] "They had no targets and we had very good targets," Dyson told an interviewer, "so if the war became nuclear we would have lost in a couple of weeks."[59]

Dyson and Weinberg both said that if they had found that a limited nuclear war was a perfectly good military option, they probably would have kept quiet about it.[60] Their report, "Tactical Nuclear Weapons in Southeast Asia," was classified Secret—though it was finally released on request through the Freedom of Information Act in 2002, after nineteen years.[61] Whether their study had any impact on the war is unclear. Secretary of Defense Robert McNamara was probably briefed on it, but he was opposed to using the weapons in Vietnam anyway.[62]

One effect was certain: a few years after it was done, the title of the report and its list of authors somehow became public, though the content and conclusions of the report did not. The academic community was largely and fervently antiwar, and the title by itself, as Dyson said, "looked pretty bad."[63] Italian and French academics wrote a letter to *Physics Today*, criticizing the Jasons for rationalizing collaboration with the military in an unjust war. Dyson's reply was printed next to their letter: "It is true that I helped write this paper under Jason auspices, and it is possible that it may have had some slight influence upon US policy in Vietnam. The question is whether I am to feel ashamed or proud of what I have done. I am glad to state publicly that I am proud of it. If my work had no effect on government policy, I can have done no great harm. If my work had some effect, I can be proud to have helped to avert a human tragedy far greater even than the one we have witnessed."[64]

TO SEE MORE SHARPLY

Dyson talked about the tactical nuclear weapons study if asked but tended not to bring it up. The study he didn't need to be asked about, the one he mentioned invariably, was adaptive optics. Adaptive optics is a complicated but useful technology: attached to a telescope, it clarifies images that the atmosphere scrambles. The military had an adaptive optics system that, beginning in the 1970s, it wanted Jason to study, to see what its future might be.

The images the military wanted unscrambled were those of the satellites Russia had been launching since the 1960s. Satellites fly high in the atmosphere, which is effectively a turbulent river that turns bright little dots into blobs that twinkle and jitter around. The military's system sensed the blob's exact distortions, then changed a flexible mirror using little pistons called actuators to reflect equal and opposite distortions, thereby cancelling them out entirely so the blob would look like a satellite.

Dyson said he set himself the question of how well such a system could do in principle, given that the atmosphere varies in milliseconds and the flexible mirror would need to change just as fast. He worked out the best way to take in the information from the image and send out instructions for moving the flexible mirror accordingly. In his words, "You have an arbitrary set of optical sensors, and an arbitrary set of mechanical actuators. Then there's a very well defined problem: to find the optimum computer program which will couple the actuators to the sensors in such a way as to give you the best possible image. I solved that problem."[65] The solution, he wrote, "turned out to be a beautiful piece of mathematics."[66]

The solution implied that the system would work only with images of things brighter than 14th magnitude—which is dimmer than the eye can see but bright enough for military telescopes looking at satellites reflecting light from the sun.[67] The work at the time was not classified; Dyson published a paper on it in 1973 in the *Journal of the Optical Society of America.*[68]

Meanwhile the military had been working on technology that would enable them to see even dimmer things. The technology measured the atmospheric distortion in the image not on the satellite itself but on a star that appeared to be next to the satellite. But of course the satellite was moving and would not always have a star next to it, so the military had been creating artificial stars by shining beams of light or lasers just ahead of where the satellite would be and measuring the distortions in the laser spot. This artificial star was called a laser guide star—"guide star" being an astronomical term for a bright star used as a reference for a nearby dim one. By now, it was the early 1980s, several years after Dyson's study, and the adaptive optics/laser system had become highly classified.

Between 1982 and 1985, Dyson worked with other Jasons on a series of secret studies, particularly on the laser guide star. One of the Jasons, a Princeton physicist named William Happer, knew that high up in the atmosphere was a layer of sodium atoms left there by vaporized meteors and that a laser tuned to make the sodium atoms fluoresce would make a bright yellow spot, like a sodium street light, called the sodium laser guide star. "So then you can actually do adaptive optics very nicely," Dyson told an interviewer, "just point the laser in the direction you want to observe."[69]

The adaptive optics system with a sodium laser guide star was of particular interest to the Reagan administration's newly declared Strategic Defense Initiative that proposed to use powerful lasers to shoot down missiles. Then the whole program disappeared completely into the classified depths of the military. Eventually the Air Force built an adaptive-optics system; it worked nicely.[70]

But Dyson and another Jason who worked on the studies, Claire Max (then at Lawrence Livermore National Laboratory), knew that dim things in the sky were interesting not only to the military but also to astronomers: "Almost everything in the universe is fainter than magnitude 14," Dyson said.[71] In fact, the idea of adaptive mirrors had come originally from astronomers.[72] Max and Dyson also knew that paired with the right telescope, the adaptive optics/sodium laser guide star system would have better resolution than the Hubble Space Telescope.[73] But neither Max nor Dyson were practicing astronomers, so the two of them visited Margaret Burbidge at the University of California at San Diego, whom Max thought of as astronomy's high priestess, and asked if astronomers would be interested; of course, Burbidge said yes. So one of the subsequent Jason reports included a chapter on the uses of the adaptive optics/sodium laser guide star system for astronomers.

But no one could tell the astronomers about such systems because the whole technology was still highly classified. It wasn't declassified until 1991, after it was clear that French astronomers were working on their own system and after Max had spent years lobbying the military in the Air Force and at Jason meetings.[74] This delay annoyed Dyson no end. "The secrecy held up progress in adaptive optics by about ten years," he said. He said that once the Air Force had an operational system, it stopped improving it; even worse, they didn't allow anyone from the outside to look at the system's effectiveness. Meanwhile, astronomers outside didn't want to spend their precious grants on possibly replicating what the Air Force was doing. "So it's a typical example of what classification does. I mean, essentially it stops progress both inside and outside," he said. "It was totally stupid."[75] Dyson, normally the soul of politeness, pronounced that word, *stupid*, with pointed clarity.

Once the system was declassified and opened up to astronomers, the new technique flourished. Currently adaptive optics systems with laser guide stars are installed on seven large telescopes with mirrors 8 to 10 meters in diameter; such techniques will also be incorporated into the upcoming extremely large telescopes with 20- to 40-meter mirrors. Jason's adaptive optics studies, said Dyson, were "great fun and was a serious science problem, and I think we did some really good work on it."

DYSON AND HOPE

Dyson was a great believer in hope.[76] He called himself an optimist, and throughout all his projects ran hope for the future.[77] And so it makes sense that he loved children; Jasons mention this unasked. One of the nicest passages in Dyson's letters home was a description of the Princeton springtime: "The children have come out of doors in summer clothes and are playing underneath the trees. No leaves are yet upon the trees, but the grass is green and the crocuses are out. The children are more brightly coloured than the crocuses."[78] Jasons remember summer sessions during which Dyson spent time talking to their children or picking up babies who had crawled off blankets. "He'd watch the kids for short periods when parents wanted to go out," said one Jason. "I thought he might be pretty old but my one-and-a-half year old loved hanging out with him."

Dyson's last summer with Jason was 2019; he died the following winter. During that last summer, he worked on studies on stockpile stewardship, on ways of keeping responses to the census confidential, and on keeping scientific research both secure and open. Jason had kept him working, even though he had long been a senior adviser rather than an active member, because his contributions were interesting and often unexpected, because of the prestige of having him there, because Jasons admired his character and mind, and because he was fun to be with. Dyson kept working almost certainly for the same reason he had joined

"I live in two worlds, the world of the warriors and the world of the victims, and I am possessed by an immodest hope that I may improve mankind's chances of escaping the horrors of nuclear holocaust if I can help these two worlds to understand and listen to each other."

in the first place: the urgent necessity of scientists and military people talking to each other.

He wrote at the beginning of *Weapons and Hope*, "I write because I live in two worlds, the world of the warriors and the world of the victims, and I am possessed by an immodest hope that I may improve mankind's chances of escaping the horrors of nuclear holocaust if I can help these two worlds to understand and listen to each other."[79] This single idea permeates his book: that the military and scientists both misuse and misestimate technology and that the rest of us are better off when the two camps communicate. With communication, Dyson thought, each camp might be a little less stupid.

NOTES

1. "Jason" is the name of the group. The group itself writes the name in all caps, JASON, as though it were an acronym. It's not an acronym; it's a reference to the Argonauts, a proper noun. I prefer to write it as though it were a name and not an acronym, if for no other reason than that the sentences are easier to read.

2. Freeman Dyson, *Weapons and Hope* (New York: Harper and Row 1984), 107.

3. Dyson, *Weapons and Hope*, 109–110.

4. Dyson, *Weapons and Hope*, 115–116.

5. Silvan Schweber, interview of Freeman Dyson, June 1998, Web of Stories 41, https://www.webofstories.com/play/freeman.dyson/41.

6. Finn Aaserud, interview of Freeman Dyson, 17 December 1986, www.aip.org /history-programs/niels-bohr-library/oral-histories/4585. On Bethe's role with the group, see Cornell Information Technologies, "About Hans Bethe," accessed 4 February 2021 at http://bethe.cornell.edu/about.html.

7. Daniel Kevles, *The Physicists: The History of a Scientific Community in Modern America*, 3rd ed. (Cambridge, MA: Harvard University Press, 1995), 337. See also Alice Kimball Smith, *A Peril and a Hope: The Scientists' Movement in America, 1945–47* (Chicago: University of Chicago Press, 1965); and Jessica Wang, *American Science in an Age of Anxiety: Scientists, Anticommunism, and the Cold War* (Chapel Hill: University of North Carolina Press, 1999).

8. David Kaiser and Benjamin Wilson, "American scientists as public citizens: 70 years of the Bulletin of the Atomic Scientists," *Bulletin of the Atomic Scientists* 71 (2015): 13–15.

9. Freeman Dyson, letter to his parents, 26 April 1962, reprinted in Dyson, *Maker of Patterns: An Autobiography through Letters* (New York: Norton, 2018), 283.

10. Dyson to his parents, 12 February 1962, in Dyson, *Maker of Patterns*, 281.

11. Dyson to his parents, 20 December 1962 ("My talent") and 16 January 1963 ("outstanding chairman"), in Dyson, *Maker of Patterns*, 287–288, 290–291.

12. Schweber, Web of Stories 112.

13. General Atomics, "TRIGA History," accessed January 2021 at https://www.ga.com/triga/history.

14. See George Dyson, *Project Orion: The True Story of the Atomic Spaceship* (New York: Holt, 2002), and chapter 5, this volume.

15. Dyson, *Maker of Patterns*, 277.

16. Freeman Dyson, *Disturbing the Universe* (New York: Harper and Row, 1979), 127.

17. Freeman Dyson, "The Future Development of Nuclear Weapons," *Foreign Affairs* 38 (1960): 457–464.

18. Dyson, *Foreign Affairs*, 460. See also Dyson, *Maker of Patterns*, 270.

19. Dyson to his parents, 7 April 1960, in Dyson, *Maker of Patterns*, 270.

20. Dyson, *Maker of Patterns*, 258.

21. John Correll, "The Neutron Bomb," *Air Force Magazine* (30 October 2017), https://www.airforcemag.com/article/the-neutron-bomb/.

22. Dyson, *Disturbing the Universe*, 129.

23. Dyson, *Disturbing the Universe*, 132. On the ACDA, see also Department of State, U. Disarmament Administration, "Arms Control and Disarmament Agency. (9/26/1961–4/1/1999)," accessed February 2021 at https://catalog.archives.gov/id/10485323.

24. Dyson, *Disturbing the Universe*, 133–134.

25. Dyson, *Disturbing the Universe*, 138–139.

26. Dyson, *Disturbing the Universe*, 139.

27. Freeman Dyson, Memorandum to Dr. Long, 31 July 1962, copy provided by George Dyson.

28. Dyson, *Maker of Patterns*, 270 ("political diatribe"); Dyson, *Disturbing the Universe*, 130 ("desperate attempt").

29. Schweber, Web of Stories 123.

30. Ann Finkbeiner, interview with Freeman Dyson, Institute for Advanced Study, 11 March 2003.

31. On the founding of the Jason group, see Finn Aaserud, "Sputnik and the 'Princeton Three': The national security laboratory that was not to be," *Historical Studies in the Physical and Biological Sciences* 25, no. 2 (1995): 185–239; and Ann Finkbeiner, *The Jasons: The Secret History of Science's Postwar Elite* (New York: Viking, 2006).

32. Freeman Dyson, private communication with author via email, 28 September 2004. See also Dyson, *Maker of Patterns*, 274, 276, 279; and Aaserud, "Sputnik and the 'Princeton Three,'" n. 203.

33. Finkbeiner, interview with Dyson, 11 March 2003.

34. Schweber, Web of Stories 123.

35. Jason members often made field trips to learn more about the subjects of their studies. In spring 1961, Dyson and other Jasons, including Nobelist Steven Weinberg, went to Key West to watch, from a destroyer, the Navy hunting submarines. Dyson wrote home that "it betrays no military secrets to say that the submarines did better than we did" (*Maker of Patterns*, 276). Weinberg spent his time talking to the crew and remembers nothing of talking to Dyson.

36. Finn Aaserud, interview of Freeman Dyson, 17 December 1986, Niels Bohr Library and Archives, American Institute of Physics, www.aip.org/history-programs/niels-bohr-library/oral-histories/4585.

37. Finkbeiner, interview with Dyson, 11 March 2003.

38. Finkbeiner, interview with Dyson, 11 March 2003.

39. Nicholas Dawidoff, "The civil heretic," *New York Times* (25 March 2009).

40. Schweber, Web of Stories 123.

41. Aaserud, interview with Dyson, 17 December 1986.

42. Schweber, Web of Stories 123.

43. All quotations in this and the next paragraph are drawn from private communications from Jason members who are allergic to public identification; the quotations come from interviews conducted by the author between October and December 2020.

44. Finn Aaserud, interview of Sidney Drell, 1 July 1986, Niels Bohr Library and Archives, American Institute of Physics, www.aip.org/history-programs/niels-bohr-library/oral-histories/4578.

45. One Jason member, asked for an interview, replied, "A warning: Freeman was the kind of person, no matter what questions you have, they're not quite the right ones"—thereby saying more about Dyson in a sentence than I can write in paragraphs.

46. Dyson, *Weapons and Hope*, vii–viii.

47. Dyson, *Weapons and Hope*, 232.

48. Dyson, *Weapons and Hope*, 275–289.

49. JASON, "Nuclear Testing: Summary and Conclusions," JSR-95-320, 3 August 1995, https://fas.org/rlg/jsr-95-320.htm. For a more critical view of "stockpile stewardship," see Hugh Gusterson, *People of the Bomb* (Berkeley: University of California Press, 2004), chap. 9.

50. Dyson, *Weapons and Hope*, 38.

51. Dyson, *Weapons and Hope*, 38.

52. Dyson, *Weapons and Hope*, 47–51, 253.

53. Finkbeiner, interview with Dyson, 11 March 2003.

54. Schweber, Web of Stories 128.

55. Nautilus Institute, *Essentially Annihilated*, "An Insider's Account: Seymour Deitchman," accessed February 2021 at https://nautilus.org/essentially-annihilated/an-insiders-account-seymour-deitchman.

56. Nautilus Institute, *Essentially Annihilated*, "Tactical nuclear weapons in 1966," accessed February 2021 at https://nautilus.org/essentially-annihilated/essentially-annihilated-tactical-nuclear-weapons-in-1966/.

57. Finkbeiner, interview with Dyson, 11 March 2003.

58. Freeman Dyson et al., Institute for Defense Analysis, Jason Division, "Tactical Nuclear Weapons in Southeast Asia," accessed February 2020 at https://fas.org

/irp/agency/dod/jason/tactical.pdf . See also Nautilus Institute, *Essentially Annihilated*, accessed February 2021 at https://nautilus.org/essentially-annihilated/.

59. Schweber, Web of Stories 128.

60. Finkbeiner, interview with Dyson, 11 March 2003, and with Steven Weinberg, 22 October 2003.

61. Dyson et al., "Tactical Nuclear Weapons," accessed January 2021 at https://fas.org/irp/agency/dod/jason/tactical.pdf; on the declassification of this report, see Nautilus Institute, *Essentially Annihilated*, "FOIA Discovery: How Nautilus Got the Story," accessed January 2021 at https://nautilus.org/essentially-annihilated/foia-discovery-how-nautilus-got-the-story/.

62. Nautilus Institute, *Essentially Annihilated*, "Insider's Account," accessed January 2021 at https://nautilus.org/essentially-annihilated/an-insiders-account-seymour-deitchman/; Sarah Bridger, "Scientists and the ethics of cold war weapons research" (PhD diss., Columbia University, 2011), 204, https://citeseerx.ist.psu.edu/viewdoc/download?doi=10.1.1.874.6959&rep=rep1&type=pdf.

63. Schweber, Web of Stories 128.

64. Freeman Dyson, letter, *Physics Today* 26 (April 1973): 13, https://physicstoday.scitation.org/doi/10.1063/1.3128011. See also Finkbeiner, *The Jasons*, chap. 5; and Sarah Bridger, *Scientists at War: The Ethics of Cold War Weapons Research* (Cambridge, MA: Harvard University Press, 2015), chap. 5.

65. Aaserud, interview with Dyson, 17 December 1986.

66. Dyson, *Maker of Patterns*, 360; Schweber, Web of Stories 126.

67. Dave Snyder, "University Lowbrow Astronomers Naked Eye Observer's Guide," September 2005, http://umich.edu/~lowbrows/guide/eye.html.

68. Freeman Dyson, "Photon noise and atmospheric noise in active optical systems," *Journal of the Optical Society of America* 65, no. 5 (1975): 551–558.

69. Schweber, Web of Stories 127.

70. Robert Duffner, "Revolutionary imaging: Air Force contributions to laser guide star adaptive optics," *Historical Perspectives* 29, no. 4 (2008): 341–345.

71. Aaserud, interview with Dyson, 17 December 1986.

72. Laird Thompson, "Adaptive optics in astronomy," *Physics Today* 47, no. 12 (1994): 24.

73. Claire Max, "Introduction to adaptive optics and its history," American Astronomical Society 197th Meeting, accessed January 2021 at http://cfao.ucolick.org/EO/Resources/History_AO_Max.pdf.

74. Horace Babcock, "Adaptive optics revisited," *Science* 249, no. 4966 (20 July 1990): 253–257.

75. Freeman Dyson, private communication with author via email, 29 September 2004.

76. Dyson, *Weapons and Hope*, 312.

77. Schweber, Web of Stories 157.

78. Dyson, *Maker of Patterns*, 148.

79. Dyson, *Weapons and Hope*, 4.

ASHUTOSH JOGALEKAR

Freeman Dyson had always been interested in biology. As a fifteen-year-old student at the six-hundred-year-old Winchester College in England, he excelled at mathematics, science, and literature. But his favorite part of being at the school was the annual school prize. He later recalled, "The best feature of the school from my point of view was the system of class prizes. In each class, three times a year, prizes were given to the winners of competitive examinations. A prize consisted of thirty shillings, at that time worth about six dollars, to be spent at the college bookstore. You could order any books you wanted within the thirty-shilling limit. I had a supply of publishers' catalogues and chose my books with care. As a result, I accumulated a small library of books stamped with the college insignia, books that I still use and treasure after 50 years, books that form the core of my intellectual development."[1]

Dyson accumulated nineteen books during his time at Winchester. These ranged from books on philosophy and literature (Ludwig Wittgenstein's *Tractatus Logico Philosophicus* and Maurice Baring's *The Oxford Book of Russian Verse*) to ones on mathematics and physics (E. T. Bell's *Men of Mathematics* and George Joos's *Theoretical Physics*). But also scattered among these were three titles by H. G. Wells. Two of them were about history and economics, and the third was about biology. Titled *The Science of Life*, the volume was written with Julian Huxley and Wells's son, G. P. Wells.[2] It is lavishly illustrated, more than fifteen hundred pages long, and contains everything there was to know about biology in the 1930s: from the smallest of bacteria to the largest of human societies. One source called it "the first modern textbook of biology."[3]

Dyson later told an interviewer what he liked best about the book: its emphasis on the details rather than some overarching philosophical theme. The structure would characterize Dyson's own life as a self-described frog, a creature that likes to solve particular problems and play around in the mud of science, in contrast to birds like Einstein that like to soar in the air and survey the grand landscape.[4] Another avenue for the young scholar's interest in the life sciences was an uncle, a doctor who was the chief medical officer of the Sudan. He was Freeman's favorite godfather and hearing his stories of solving the medical problems of a large country got Freeman interested in wanting to become a doctor himself.

But it was not to be. Freeman's biological interests and talents were orthogonal to each other.[5] This point was driven home when he brought home some crayfish to dissect, knowledge of anatomy being a skill that seemed imperative to master if he wanted to become a doctor. Sadly, the experiment was hopeless, and he had trouble finding even the basic parts of crayfish anatomy. Meanwhile, the overriding interest from Freeman's boyhood had been mathematics. Thereafter, even as he became one of the most distinguished mathematical physicists of the twentieth century, Dyson retained a consuming interest in biology throughout his career. He befriended biologists, wrote about biology in elegant prose, gave testimony on biology to Congress, tackled several biological problems as part of a government advisory board called the Jasons (described in chapter 6), and pondered the fate of human biology in outer space.[6] And in the form of the one serious technical contribution he made to biology, he came up with a novel model for the origins of life, a puzzle that is as grand as any in science.

Dyson worked for Bomber Command as a statistician during the Second World War and graduated with a degree in mathematics from Cambridge University in 1945. At that time, biology was in the throes of a revolution, although most of the pioneers of this revolution themselves did not know that. In 1946, Dyson had a chance encounter with Francis Crick before he gained fame as one of the leading figures of modern molecular biology. When Dyson met Crick, Crick had just finished a depressing stint working on underwater mines for the British Admiralty.[7] At thirty-three, Crick felt he was too old to contribute to physics. At King's College and Cambridge, physicists like John Randall and Lawrence Bragg were establishing fledgling biology laboratories that would gaze deep into the structure of biological molecules, especially proteins. Dyson remembered Crick telling him how he intended to switch from physics to biology because biology looked exciting. "No, you're wrong," he said. "In the long run biology will be more exciting, but not yet. The next 20 years will still belong to physics. If you switch to biology now, you will be too old to do the exciting stuff when biology finally takes off."[8] Wisely, as Dyson himself admitted later, Crick did not heed his advice. Seven years later, in 1953, Crick and a young James Watson cracked the structure of DNA.[9] The molecular biology revolution was underway.

After the war, Dyson retained a lively interest in biology, even as he moved to the United States and delved deep into theoretical physics, nuclear reactor design, and nuclear propulsion.[10] In 1947, for example, while he was traveling throughout the United States, he had an opportunity to witness Melvin Calvin's experiments on deciphering the path of carbon through photosynthesis using radioactively labeled carbon. Dyson was so excited by this work that he gushed with excitement in a letter to his parents, perhaps the only letter among the copious ones he sent to his parents that delved into so much scientific detail: "Today I saw a scientific miracle which will turn the world upside down; perhaps even save lives . . . it will do for biology what the Wilson cloud chamber did for physics."[11] The reference to the cloud chamber, an instrument that opened up the world of particles to physicists by allowing them to track their paths in a mist of fine droplets, is especially interesting. It is clear that even at this point, Dyson was appreciating the impact that new techniques and instruments could have on scientific revolutions. He was to write often about these themes later.[12]

The 1960s were a particularly exciting era for biology. After the structure of DNA had been discovered, the next big challenge was to figure out how DNA encoded the proteins that were the workhorses of life. A large number of biologists, including physicists-turned-biologists like Crick, Max Delbrück, and Leo Szilard, played a role in this endeavor. By the end of the decade, not only had the genetic code been cracked and the role of RNA discovered, but the basics of the mechanism of gene regulation had also been worked out.[13] Most tantalizing, the first steps toward deciphering and synthesizing short sequences of genes had been taken. It was also in the 1960s that Dyson developed a serious technical interest in biology, one that was to lead to a major contribution to thinking about the origins of life.

ORIGINS OF THE SEARCH FOR THE ORIGINS OF LIFE

Along with the origin of the universe and the origin of consciousness, the origin of life is among science's grandest questions. The problem is a difficult one, so difficult that Charles Darwin himself set it aside in his great work, *The Origin of Species*. In that work Darwin laid out a mechanism for how life evolves once it gets started, but said nothing about how it began in the first place. However, he did leave a tantalizing and memorable comment on the topic to his friend Joseph Hooker in a 1871 letter:

It is often said that all the conditions for the first production of a living organism are now present, which could ever have been present.—But if (& oh what a big if) we could conceive in some warm little pond with all sorts of ammonia & phosphoric salts,—light, heat, electricity &c present, that a protein compound was chemically formed, ready to undergo still more complex changes, at the present day such matter would be instantly devoured, or absorbed, which would not have been the case before living creatures were formed.[14]

The "warm little pond" metaphor would persist, to both fruitful and misleading ends. Research on the origins of life did not become a respectable field of inquiry for the next several decades, although tendrils of it can be seen in developments pertaining to the work of Louis Pasteur, Lazzaro Spallanzani, Joseph Lister, and others definitively establishing the presence of microorganisms as the cause of disease and invalidating the doctrine of vitalism.[15] Also particularly relevant to research in the field was Eduard Buchner's discovery that living cells were not required for the process of fermentation; instead, their extracts, which contained molecular entities—enzymes—would do. Buchner received the Nobel Prize for this work and became a leading figure of biochemistry, demonstrating that mere molecules can perform the myriad, wonderful functions that were previously considered to be the exclusive preserve of life.[16]

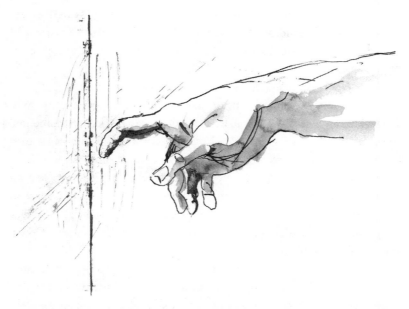

"It is often said that all the conditions for the first production of a living organism are now present, which could ever have been present."

A serious foray into the origins of life began with Russian biochemist Alexander Oparin's work in the 1920s. In retrospect, scholars have recognized this work was revolutionary; not only did Oparin propose a general blueprint for life's origins, but he also addressed the critical thermodynamics questions that remain at the heart of the problem. Oparin proposed that when life arose, it could not make its own food and had to acquire it from other life. He further proposed that life could potentially contravene the second law of thermodynamics by constantly reducing disorder through constructive food consumption. As a simple model of life, he proposed a class of molecular structures called "coacervates." A coacervate consists of a liquid phase that can separate from a colloidal solution by addition of another liquid phase.[17] In one form or another, all of these ideas were to persist as important components of ideas regarding life's origins. Unfortunately, Oparin became associated with the discredited Soviet biologist Trofim Lysenko, leading Western scientists to discard his ideas. After Stalin's death in 1953, both Oparin and Lysenko were ousted from their influential positions in the Soviet scientific hierarchy.[18]

Two big conceptual steps were taken in the 1940s and 1950s, steps that while not dwelling directly on the origins of life, provided thought-provoking blueprints for how to think about life as a distinct phenomenon. The first development was the publication in 1945 of a series of lectures given in 1944 by the physicist Erwin Schrödinger at Trinity College in Dublin. Dublin was then an oasis of peace, far removed from the war tearing Europe apart. By that time, Schrödinger had already secured his place in history as one of the founders of quantum mechanics, so he could afford to think about areas outside his field of expertise and, in his own words, risk making a fool out of himself.

Schrödinger contributed two important ideas regarding how to think about life. One was to postulate that the genetic material was an "aperiodic crystal" that would perpetuate itself by serving as a template for other similar crystals. The second, echoing Oparin, was to suggest that the way life persists despite the second law of thermodynamics is by extracting "negative entropy" from the environment, using energy to create order in a universe in which things tend toward disorder. Schrödinger's lectures were published as a little book of ninety-one pages titled *What Is Life?*[19] It remains one of the classics of twentieth-century science, succinctly and provocatively written with a clarity that should be a role model for science writing. It served as inspiration for Francis Crick, James Watson, Sydney Brenner, and many other pioneers of molecular biology, who became convinced that the nature of the gene could be found by thinking about it systematically, the way Schrödinger did. Although few of Schrödinger's hypotheses about the origins of life have withstood scrutiny, his little book clearly articulated potent questions and inspired a new generation to take them seriously.

The other contribution to thinking about life, underappreciated at the time, was a set of lectures given by the mathematical physicist John von Neumann in 1948 at Caltech. These were titled "The General and Logical Theory of Automata" and published in 1951.[20] With a mathematician's precision and attention to generality, von Neumann laid out in the abstract several of the essential structures that would be needed to replicate life. For instance, he described components like a blueprint (specifying the instructions for making new genetic material), a factory (that actually constructs new genes), a controller (that regulates what the factory makes), and a duplicating machine (that makes a copy of the genetic material). It is striking that these components broadly correspond to the DNA, RNA, ribosomes, and replicase enzymes that were identified only much later. Yet von Neumann's work largely remained in the shadows of the biological stage.[21] As the science writer John Casti puts it, "Such are the wages of the theoretician, especially one who solves 'only' the general case!"[22] Also underappreciated about von Neumann's ideas was that they clearly separated the instructions for making something from the mechanism that actually makes it. This distinction was consonant with the foundations of computing that von Neumann was working on and pertained to what we today call "software" and "hardware." The distinction is important not only in computer science but in life's origins and its perpetuation.

While Schrödinger's and von Neumann's ideas were important for understanding the question and the potential answer, tangible progress could come only from experimental work, and—since the origin of life necessarily involved the origin of molecules—especially from chemistry. The next big step indeed came from a chemist in 1953, incidentally the same year that Watson and Crick published the structure of DNA and Fred Sanger published the structure of proteins. Goaded by a suggestion by physical chemist Harold Urey, Stanley Miller simulated Darwin's "warm little pond" by boiling hot water in the presence of sparks and a so-called reducing atmosphere that consisted of methane, ammonia, and hydrogen. To Miller's surprise, the resulting solution turned out to have a small concentration of the critical amino acids that make up proteins.[23]

Stanley Miller works on his 1953 experiment at the University of Chicago that demonstrated that amino acids could be formed under conditions comparable to those present in the Earth's atmosphere early in our planet's history. Source: Getty Images.

Miller did not prove that this was how life started; rather, his experiment demonstrated that the molecules of life could be created by simple physical processes. Over the next few decades, many variants of the Miller-Urey experiment have been run under different reaction conditions, and each time researchers have observed a different mix of molecules. However, getting such experiments to synthesize a high concentration of molecules, in particular DNA and RNA, has proven elusive.

DYSON'S CONTRIBUTIONS: METABOLISM FIRST

The preceding discussion sets the stage for Dyson's work in the area. In 1964–1965, Dyson was teaching physics at the University of California in San Diego during a sabbatical from the Institute for Advanced Study. He had been to San Diego twice before, once to work on an intrinsically safe nuclear reactor, the TRIGA (Training, Research, Isotopes, General Atomics) and again to work on a spaceship powered by nuclear bombs (described in chapter 5). While teaching at the university, he used to walk across the street to see a friend at the Salk Institute of Biological Studies, Leslie Orgel. Orgel was a chemist famed for his eponymous "Orgel's Rules," one of which says that "evolution is cleverer than you are."[24] He was a versatile chemist who had started out in inorganic chemistry and crystallography—he had been one of the first to see Watson and Crick's new model of DNA in 1953—and later moved into biochemistry. He was particularly interested in knowing how a set of molecules can self-replicate. Dyson credited Orgel with teaching him most of the biochemistry he knew and for encouraging him to think about the origins of life.[25] Orgel's main achievement was to imagine plausible conditions that would allow RNA to form and polymerize.[26] Even then it was known that whereas DNA is the blueprint for the genetic material, RNA does most of the work in encoding proteins, the workhorses of life. Orgel came up with the idea of an RNA world in which RNA formed before DNA and proteins. Two decades later, his hypothesis would receive a major boost when it was found that RNA itself can also serve as an enzyme. Ironically,

Orgel was to strongly disagree with Dyson's ideas regarding the origins of life.

Dyson kept on thinking about his discussions with Orgel over the next few decades, even while he continued his work in mainstream physics, arms control, and science writing. In September 1981, he attended Matter into Life, a conference in Cambridge, England, organized by astronomer Martin Rees, at which he talked about his ideas. A paper duly followed in 1982, his first technical contribution to the field. Titled "A Model for the Origin of Life," it was published in the *Journal of Molecular Evolution*.[27] But the real opportunity to expound on his ideas for a general audience came in 1985 in the form of the Tarner Lectures. The lecture series started at Cambridge University in 1916 to address "the Philosophy of the Sciences and the Relations or Want of Relations between the different Departments of Knowledge." Previous lecturers had included Bertrand Russell, Schrödinger, and E. O. Wilson; the series was meant to cross disciplinary boundaries. Dyson's Tarner Lectures were published in 1985 under the title *Origins of Life*; a second edition appeared in 1999.[28] Similar to Schrödinger's little book of ninety-one pages, Dyson's book was a mere one hundred pages. And just like Schrödinger's book, Dyson's was written with enviable clarity.

Dyson's account starts by acknowledging his illustrious forebears who seem to be roughly divided among physicists, chemists, and biologists. The physicists include Schrödinger and von Neumann, the chemists feature Orgel and Manfred Eigen, and the biologists are Lynn Margulis and Motoo Kimura. Eigen demonstrated the remarkable efficiency with which RNA can replicate in a test tube. Margulis proselytized the pioneering idea of symbiosis, which is responsible for some of the existential milestones in the evolution of life. Kimura came up with the very important idea of random genetic drift and developed the mathematics for it. All of them deeply inspired Dyson, some because he agreed with them and others because he disagreed.

In his lectures, Dyson professes no expertise on the origins of life and especially professes ignorance of chemistry, but like Schrödinger and von

Neumann, he explains that his main purpose is to bring a physicist's point of view to studying the problem. He is the quintessential outsider approaching a problem from a novel angle. His basic premise starts from recognizing that Schrödinger and von Neumann did not ask all the questions that needed to be asked. Most notably, they focused on only one of the two main processes responsible for life: replication. The other important process is metabolism. If one thinks about what makes up the essence of being human, for instance, two aspects seem paramount: the first is the capacity of our cells to replicate and of our bodies to reproduce, and the second is our capacity to breathe, to consume and use energy from food, and to think with our unusually large brains. The first two features are examples of replication, the others of metabolism. But Schrödinger and von Neumann, as well as Orgel and Eigen, focused only on replication. Their view of the magic elixir that got life started was a world in which molecules could replicate, by themselves or with the aid of another. Replication was the essence of life for them because it was the only way to transmit hereditary information from one generation to the next. DNA is the canonical example of a replicator, with the exact sequence of its base pairs in a parent cell faithfully copied to another strand of DNA in a daughter cell. They considered metabolism to be a logical consequence of replication and therefore not a separate process. Dyson strongly disagreed.

Dyson's model of the origins of life starts with the hypothesis of metabolism without replication and in fact suggests that metabolism might have come before replication. In computer terms, it's like having hardware without software, a notion that's novel and important precisely because of its implausibility.[29] Dyson's theory is therefore one of a few scenarios for the origin of life that go under the heading "metabolism first." Another way to say it is to postulate that life arose not just once, with replication and metabolism intimately intertwined, but twice or even multiple times, with the two functions arising independently. This is called the dual-origin scenario.

Why should one take seriously the hypothesis that metabolism might arise independently from replication, especially since—to this day—no one has identified viable biological organisms that can metabolize without replication? The key problem with replication is that it needs to be highly accurate. Orgel's and Eigen's work showed that unless it is so, errors introduced by imperfect copying and mutations quickly accumulate in each replication cycle.[30] Schrödinger had also recognized this challenge and wondered how life could reproduce so faithfully if errors accumulated. It is hard to imagine how life could have gotten started with molecules that had the kind of accuracy necessary for being good replicators. At the very least, one would need to postulate a mechanism by which such molecules were created and evolved. Currently the finest examples of known replicators are DNA and RNA, but both are complicated molecules, consisting of three independent parts (the base, the sugar, and the phosphate), and there is no definitive evidence that these molecules could have arisen by themselves as the first progenitors of life, let alone that they could have catalyzed their own replication in a self-sustaining manner. In addition, even today, DNA and RNA are prone to a high degree of error while replicating, and the only thing that prevents life from descending into a runaway catastrophe of errors is the presence of error-correcting enzymes, enzymes that could not have existed in the beginning.

Dyson counters that the key is to realize that "molecules replicate, but populations of molecules and cells can reproduce and metabolize."[31] Unlike exact replication, reproduction and metabolism are statistical phenomena that can be inaccurate yet error tolerant. Also, unlike replication, metabolism can be a messy process that does not depend on pristine conditions. During the early history of life, when Earth was a complicated, ever-changing environment with violent conditions like volcanic eruptions and abrupt climate change, it was much more likely for life to reproduce inaccurately than to replicate accurately. As Dyson points out, the German word for metabolism is *Stoffwechsel*, which can be roughly translated as "stuff exchange."[32] The laws of chemistry dictate that when two molecules interact, there is a good chance that the interaction will lead to some stuff from one molecule sticking to another. Thus, what Dyson calls "metabolism" should be understood as a much more primitive process of chemical exchange and therefore much more likely, especially on the early Earth.

The question is whether a population of molecules can reproduce accurately enough so that the information they encode can be propagated to the next generation in a self-sustaining manner. To explore this question, Dyson builds what he calls a "toy model" of a kind well known to physicists: it aims to identify what kind of accuracy and molecular populations would be needed to transition from a state of disorder to one of order. To Dyson, this transition in essence is the definition of the origin of life. In the preface to his book, Dyson writes that the two main features of his model are error tolerance and diversity, qualities that could be said to mark his own career.

A model is a simplified, often idealized picture of reality that should make testable predictions. Physicists since the time of Galileo have become accustomed to imagining simple models for complicated physical phenomena. Biologists instead tended to start with the hard work of collecting large quantities of data before turning to model building, which really took off with the advent of population genetics and molecular biology. The Watson and Crick model, in this sense, can be said to be a particularly good example of an early model of a biological molecule, omitting many complications but retaining the key aspects of complementary base pairing and replication. Just like models in physics, models in biology can be powerful aids in stripping away the complications of real biological processes and reducing them to simple, testable "virtual experiments" that can be performed for a fraction of the cost and time.[33] Especially in the context of origins of life, models of the kind Dyson built are reminiscent of some of the first models of artificial life built by pioneers like Nils Barricelli and Tom Ray.[34]

To build his model, Dyson draws on the work of the population geneticist Motoo Kimura and the biologist Ursula Niesert. Kimura pursued the idea that random drift could explain evolution in large populations.[35] Niesert developed a theoretical model of evolution that tries to understand the effects of the competing processes of fluctuation, mutation, and death on the transition between disorder and a viable population of self-sustaining, ordered molecules.[36] Dyson applies the same mathematics that Kimura applies to populations of organisms to populations of molecules or cells. Like his illustrious predecessor von Neumann, Dyson has the advantage of being a generalist in this case, looking at the abstract, general case and leaving the details to be worked out by scientists doing experiments with real molecules; his work is thus an invitation to experiments that can prove or disprove his model. Like all other good models, his makes a number of simplifying physical and mathematical assumptions; among these are the assumption of metabolism without replication and random drift without Darwinian natural selection.

In Dyson's model, a population of molecular components called monomers—for instance, the individual amino acids that comprise proteins—can combine using chemical catalysis reactions into complete molecules. Like Niesert's model, the model can transition from a completely disordered state in which molecules don't do anything interesting to a completely ordered state in which molecules carry out the chemical functions of life. Dyson calls the first state "hot sulfide soup," signifying death, and the second state "Garden of Eden" or "immortal," signifying life. The most interesting region is the one between these two because it signifies the properties necessary to jump from sterility to life. Another analogy that Dyson borrows is the Curie-Weiss model of phase transitions, which models the transition between magnetic order and disorder in a ferromagnet.[37]

Dyson's model has three key parameters: a, the number of monomers that can assemble into polymers; b, the number of kinds of catalytic reactions that can achieve this assembly; and N, the size of the molecular population resulting from this assembly. Dyson plays with several values for the three parameters and, crucially, finds a rather limited range of values for which the model will show a sustainable transition to a viable assembly of polymers rather than dissolving into a dead soup of monomers without activity. These values are a ranging from 8 to 10; b ranging from 60 to 100; and N ranging from 2,000 to 20,000. The range for a

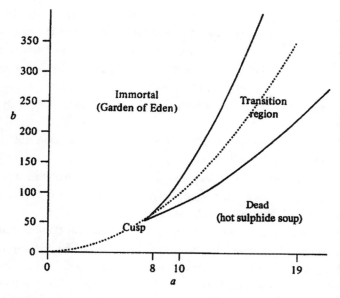

Key features of Dyson's model for the origins of life, including two regions labeled "Immortal" and "Dead," with a critical transition region between them. In Dyson's model, a represents the number of monomers that can assemble into polymers, and b represents the number of kinds of catalytic reactions that can achieve this assembly. Reproduced with permission from Freeman J. Dyson, *Origins of Life*, 2nd ed. (New York: Cambridge University Press, 1999 [1985]), 65.

is particularly interesting because we know that twenty amino acids are needed for proteins to function. The model definitely fails with $a = 4$. We know that four bases are needed for modern RNA and DNA to function, yet Dyson's model suggests that this number might be inadequate to ensure an orderly transition. Dyson does not propose that the first metabolizing molecules might have been proteins; rather, he says that his model is consistent with what we know about the number of monomers needed to form functioning proteins.

But because metabolism is messy, any model that argues for a metabolic origin for life must be able to tolerate a high error rate; replication, however, cannot tolerate a high error rate. The biggest advantage of Dyson's model is that it allows imperfect yet persistent reproduction even with error rates of 20 to 30 percent. For instance, an error rate of 25 percent in the context of DNA would mean that one out of every four bases is installed incorrectly during replication. This error rate would be catastrophic for replication, but does not threaten the far simpler transition to order in Dyson's model.

The metabolism-first scenario clearly posits that replication and metabolism can be decoupled. But clearly replication had to kick in at some point to account for life's propagation. How did the first metabolizing entities acquire the all-important capacity for replication? The solution Dyson proposes is based on Lynn Margulis's theory of endosymbiosis. Margulis was a scientific pioneer who identified the evolution of two organelles without which life would not be possible: the mitochondrion and the chloroplast.[38] Her theory was first proposed by Russian scientists like Konstantin Mereschkowski in the early 1900s but then fell by the wayside. When Margulis tried to publish her idea in the 1960s, it was rejected by fifteen scientific journals before being published in the *Journal of Theoretical Biology*.[39] Today, the endosymbiosis hypothesis is regarded as one of the most important ideas in biology and is the leading theory for the evolution of complex, eukaryotic cells from simple, prokaryotic cells.[40]

Mitochondria are responsible for the process of oxidative phosphorylation that generates energy.[41] Chloroplasts are responsible for photosynthesis that provides life-giving oxygen. Together, mitochondria and chloroplasts enable complex, multicellular life on this planet. In Margulis's theory of endosymbiosis, mitochondria were formed when a low-energy, multicellular, eukaryotic organism randomly swallowed a high-energy, prokaryotic, single-celled organism. In time the two learned to live in peace, the bacterium using the prokaryote's energy-generating apparatus to fuel metabolic processes that needed high energy and the prokaryote depending on the bacterium's resources to sustain itself. The distinction between the two was eventually erased; the union became complete, and multicellular life got established. A similar process took place with the chloroplast, in which a blue-green alga that could photosynthesize was swallowed by a primitive bacterium that could not. In both cases, the chloroplast and the mitochondrion were a kind of parasite that "infected" the other cell but soon adapted to live in harmony with it.

Dyson believed that since endosymbiosis had been shown to be so crucially important for two of life's most fundamental processes, it could very well be responsible for the fusion between metabolism and replication that led to an inaccurate but more plausible process combining with an accurate but implausible process, leading to a winning recipe for life. He left the door open in terms of the exact identity of both the metabolizer and the replicator, as he did in terms of the basic metabolizing entities in his model.[42]

In sum, Dyson's deceptively simple model postulates that metabolism came before replication, because metabolism—unlike replication—does not need to be accurate; the model has sustainable inaccuracy built into it. It is testable and makes quantitative predictions about the number of monomers needed to sustain the transition of a population of molecules from order to disorder. It is agnostic about the exact identity of these molecules, but suggests that they could be more protein-like rather than DNA-like. It proposes that replication started out as a parasitic process on metabolism. It remains an exemplary model of a metabolism-first, dual-origin scenario for the origins of life.

MORE RECENT DEVELOPMENTS

Since the 1980s there have been several significant advances in the study of the origins of life. Both on the metabolism-first and the replication-first fronts, new findings and ideas have emerged. The resulting picture is a complex one, sometimes speaking to the importance of metabolism and sometimes to that of replication.

Stanley Miller's original model of a warm soup of molecules has been strongly challenged, partly on the basis of new findings that indicate a nonreducing atmosphere on early Earth and partly because the model does not address how factors like dilution and environmental changes could keep the molecules of life from assembling together.[43] Nonetheless, Miller's general theme of having some kind of physical process create the molecules of life still commands attention among experts in the field.

The replication-first model of life has seen some spectacular advances since Dyson proposed his alternative model. The most important finding was the discovery that RNA can catalyze its own breakage and formation, as well as that of other nucleic acids like DNA.[44] This self-splicing RNA led to the idea of an RNA world, one in which RNA was a favored replicator because it could perform the functions of both replication and at least one kind of metabolism.[45] Even today, the RNA world remains the most-favored theme for the origins of life. However, even RNA proponents have had a hard time postulating how exactly the first self-splicing RNA might have arisen on Earth and how it could have performed its functions under prebiotic conditions. An important advance in this area has been the work of John Sutherland from the Medical Research Council (MRC) Laboratory of Molecular Biology in Cambridge, England, who in a series of papers showed that the three parts of RNA—the ribose, the base, and the phosphate—could creatively assemble in nontraditional ways starting from simple chemicals like cyanide.[46] But Sutherland's work also relies on the existence of a concentrated source of molecules—water will generally tear any such fragile precursors apart. Sutherland and others have proposed that one set of conditions that might circumvent this problem is wet-dry cycles that may have occurred on Mars during its evolution. The NASA Perseverance rover mission to Mars aims to test this hypothesis.[47]

In many ways, the most important new advance in the field, and one fully consistent with Dyson's metabolism-first approach, has been a focus on energy.[48] Both Dyson and Margulis were on the right track when they pointed to the example of energy-producing mitochondria as parasites on primitive reproducing cells. Stretching this idea more generally, whether metabolism or replication came first, both processes need energy. The search for plausible sources of energy thus became an important part of scientists' thinking. Starting in 1977, a series of remarkable explorations led to the discovery of hot hydrothermal vents deep in the ocean. Some of these were acidic and others were alkaline. A remarkable diversity of life teemed around them. The difference between acidic and alkaline vents is crucial and was first recognized by Mike Russell and Bill Martin. Building on a forgotten idea by a French doctor named Stéphane Leduc who argued—in a prescient anticipation of metabolism-first models— that "the essential phenomenon of life is nutrition," Russell in particular was drawn to metabolism.[49] The important clue for him was the chemiosmotic theory, first postulated by the British biochemist Peter Mitchell, which proposed that the essential energy-generating apparatus in complex cells is a proton gradient, so named because it uses positively charged protons working against a negatively charged gradient to produce adenosine triphosphate (ATP), the energy currency of life. If the proton gradient is to work, then the environment that gave rise to life had to be alkaline rather than acidic, since acids donate protons whereas alkaline bases accept them. Russell was also inspired by the ideas of the patent-attorney-turned-chemist Gunther Wächterhauser, who thought that iron sulfide (called pyrite) would have been conducive as a primordial material for the first living organisms. In hot vents, hydrogen can be generated from reacting iron with hydrogen sulfide, a gas that is present in the vents and can provide a source of electrons that can create simple chemical precursors for complex compounds like amino acids and nucleic acids.[50] In 2000, such alkaline vents were discovered in the Atlantic Ocean, producing simple gases like methane and hydrogen, which could be a source of energy.[51] Whether these vents could indeed give rise to life-like entities is an open question.

There is also a third, related albeit distinct, alternative that has emerged: that of compartments. Whether nucleic acids or proteins came first, there had to be some kind of compartment in which they could sequester themselves to be protected from the environment and grow. These compartments are now called "protocells."[52] The prime candidates for protocells are lipids; the cell membrane in our cells is made out of phospholipids, and lipids are the only molecular structures we know that can self-assemble easily without external agents into a compartment. Since the early 2000s, the biochemist Jack Szostak has demonstrated in major studies not only how lipids can form protocells and entrap pieces of RNA inside them, but also how this RNA can copy itself if supplemented with simple chemicals like citrate and magnesium ions. Szostak and his team also showed how these protocells could divide and "reproduce," carrying the RNA along. Not only that, but protocells with RNA seemed to preferentially divide without disruption, showing a primitive kind of Darwinian evolution.[53]

Another promising direction has been taken by Doron Lancet at the Weizmann Institute in Israel, whose work Dyson references in *Origins of Life*. Lancet has carried out simulations of lipid protocells in which the lipids not only divide but also catalyze their own formation.[54] Starting from a mixture of lipids, Lancet has shown that lipids that can catalyze their own formation better than others preferentially divide and form more protocells, catalyzing even more of their own species in a process of positive feedback and Darwinian selection. Lancet's lipids are so self-sustaining that he has called the origins of life the "lipid world," akin to the RNA world.[55] A geneticist whose main area of work is the process of olfaction, Lancet says that he is an outsider just like Dyson and was inspired to get into the origins of life field in the 1980s after reading Dyson's book.[56]

This work indicates that whether metabolism or replication came first is a complicated question, partly because, as the work by Szostak and Lancet shows, it is possible to imagine a hybrid system that does both—a system that involves self-reproducing molecular entities housing other self-replicating ones. But generally the metabolism-first viewpoint that Dyson championed remains an odd combination of both a contribution from a maverick minority and an implicit part of the latest research in the field. In some sense, it is becoming subsumed in more recent work; its major theme of populations of molecules reproducing and becoming self-sustaining has become part and parcel of the field.

ENGINEERING GENES AND CLIMBING TREES

One of the most interesting qualities of Dyson's model, which he discusses at some length in the last few chapters of *Origins of Life*, is the disordered state that he calls "death." Why can't a population of cells remain forever in an ordered state of life, a veritable immortal Garden of Eden? Dyson's model allows only random drift, not Darwinian evolution. Once Darwinian evolution is introduced, then some individuals with better fitness will necessarily weed out other individuals with worse fitness in an environment of limited resources. Death is not optional if Darwinian selection is to operate. As Dyson puts it, "Life had to invent death in order to evolve."[57] No model of the origin of life can be a true model if it lacks a concept of death, the winnowing out of the old and proliferation of the new; no proliferation of the new can take place until we can use the implicit knowledge created by the old.

Although Darwinian selection and random genetic drift were essential for the evolution of the first molecules of life, we are now in an age in which humans can artificially engineer DNA and RNA at a molecular level. After the structure of DNA and RNA and the genetic code were worked out, we acquired the capability to read, copy, edit, and potentially write genomes ourselves. These developments hold great promise in curing disease but can also turn humans into hideous creatures, as H. G. Wells recognized more than a century ago.[58] At the very least, the benefits of genetics research might accrue unequally, benefiting rich more than poor and giving them an unfair advantage. In 1975, at the Asilomar conference, life scientists debated the pros and cons of allowing scientists to create organisms with recombinant DNA.[59] The question soon exploded into public controversy, leading to heated debates among research scientists, local officials, and concerned citizens across the United States.[60] Dyson participated in the Princeton citizens' council. Not only did he issue a statement summarizing his opinion on the Princeton committee's deliberation, but he also testified before Congress on the pros and cons of genetic engineering in 1977.[61]

Dyson's main argument was that whereas the ethical and scientific harm potentially enabled by scientific technology was very real, he would rather err on the side of allowing the research. No one could predict whether a particular piece of scientific research could be used for harmful purposes, he insisted. To illustrate his point, Dyson gave the example of John Milton's speech before the English Parliament in 1644 arguing for the unlicensed printing of books. It was much more dangerous for intellectuals to publish their views in 1644 compared to 1977, but Milton put up a vigorous defense in favor of allowing books to be published and perhaps restricting them only after publication. Just like Milton, Dyson was of the opinion that because science is fundamentally unpredictable, its fruits can be regulated only after they ripen. He voiced similar thoughts in a debate with the computer pioneer Bill Joy in the 1990s and predicted that the time had come when biology as a discipline would be domesticated.[62] His vision of genetic engineering's future was one in which, just like the computer pioneers, young people would build exciting biological products in their garages and children would engineer fun biological organisms from cheap kits.[63]

Dyson continued to work on unconventional ideas about biology for the rest of his life; as late as 2012, he coauthored with Bill Press a major paper challenging game-theoretic assumptions in evolutionary biology that changed thinking in the field.[64] One of his main interests was the future of life in space. He proposed looking for frozen fish in the orbit of Europa, since that would be much easier—even if more improbable—than digging deep for life under Europa's ice-covered ocean.[65] He imagined sending self-replicating probes (inspired by John von Neumann) that would copy and build themselves out of water and minerals plentifully found in the solar system.[66] He thought that one could plant giant trees—perhaps with genetically engineered leaves that could efficiently photosynthesize—on comets, with the low gravity enabling them to grow hundreds of feet tall. In 2001, he gave a talk arguing that life is as much analog as digital and that one needs to understand analog computing in the brain to develop true artificial intelligence.[67]

In the report that Dyson wrote on the Princeton committee's deliberations on the use of recombinant DNA technology, he compared the process of allowing potentially harmful research to escape out into the world to allowing children to climb trees. Parents often don't allow their children to climb trees because a fall might have very serious consequences. But "a wise parent also knows that, no matter how terrible the consequences of an accident may be, we cannot always be saying 'no.' If we forbid tree-climbing arbitrarily and absolutely, we not only destroy a part of the joy of childhood but also destroy the respect upon which our authority as parents ultimately depends."[68] Freeman Dyson loved to climb. When he was an undergraduate at Cambridge, he climbed the tall university spires (as described in chapter 1). And ever since he was a child, he loved climbing trees, retaining the habit even after becoming a professor. From that unique vantage point atop his beloved trees—just as Newton likened himself to a small boy playing on the seashore—Dyson saw a bit farther than the rest of us.

NOTES

1. Freeman Dyson, "Comments on selected papers," in Dyson, *Selected Papers of Freeman Dyson with Commentary* (Providence, RI: American Mathematical Society, 1996), 1–49, on 3.

2. H. G. Wells, J. Huxley, and G. P. Wells, *The Science of Life* (London: Cassell, 1931).

3. David C. Smith, *H. G. Wells: Desperately Mortal* (New Haven, CT: Yale University Press, 1986), 263.

4. Freeman Dyson, "Birds and frogs," *Notices of the American Mathematical Society* 56, no. 2 (2009): 212–223.

5. Freeman Dyson interview with Silvan Schweber (1999), Web of Stories, https://youtu.be/fE3hOO49-CU.

6. On Dyson's participation in the JASON group, see chapter 6, this volume.

7. Matt Ridley, *Francis Crick: Discovery of the Genetic Code* (New York: Harper Perennial, 2009).

8. Freeman Dyson, *A Many-Colored Glass: Reflections on the Place of Life in the Universe* (Charlottesville: University of Virginia Press, 2007).

9. Years later, Dyson was to borrow a page from Watson when the latter's book, *The Double Helix* (New York: Atheneum, 1968), came out. Dyson marveled at how Watson could remember so much detail from his time in Cambridge. Watson told Dyson it was possible because he had kept all the letters he sent to his parents. Dyson decided to do the same, and it made it possible for him to write *Maker of Patterns: An Autobiography through Letters* (New York: Norton, 2018).

10. On Dyson's contributions to quantum electrodynamics, see chapter 3, this volume; on nuclear reactors and nuclear propulsion, see chapter 5, this volume.

11. Freeman Dyson to his parents, 26 August 1948, reproduced in Dyson, *Maker of Patterns*, 95.

12. Freeman Dyson, *The Sun, the Genome and the Internet: Tools of the Scientific Revolution* (New York: New York Public Library, 1999). On the impact of cloud chambers on particle physics, see Peter Galison, *Image and Logic: A Material Culture of Microphysics* (Chicago: University of Chicago Press, 1997), chap. 2.

13. Horace Freeland Judson, *The Eighth Day of Creation: Makers of the Revolution in Biology* (New York: Touchstone Books, 1979).

14. Francis Darwin, *The Life and Letters of Charles Darwin* (London: John Murray, 1887), 202–203.

15. Paul DeKruif, *Microbe Hunters* (New York: Mariner Books, 2002 [1926]).

16. Ulf Lagerkvist, *Enigma of Ferment: From the Philosopher's Stones to the First Biochemical Nobel Prize* (Singapore: World Scientific, 2005).

17. Quite interestingly, Oparin's strictly academic idea for the origins of life has now become the basis for a view of certain diseases as being caused by liquid-liquid phase transitions inside cells. Studying and potentially disrupting these so-called condensates with drugs is the basis of a recent biotechnology company (https://dewpointx.com/). Note: The author does not have financial or other interests in this organization.

18. On Lysenkoism and responses by scientists in Western Europe and the United States, see David Joravsky, *The Lysenko Affair* (Chicago: University of Chicago Press, 1970); Loren Graham, *Science in Russia and the Soviet Union* (New York: Cambridge University Press, 1993), chap. 6; and Michael Gordin, "Lysenko unemployed: Soviet genetics after the aftermath," *Isis* 109 (2018): 56–78.

19. Erwsin Schrödinger, *What Is Life? The Physical Aspect of the Living Cell* (Cambridge: Cambridge University Press, 1945).

20. J. von Neumann, "General and logical theory of automata," in *Cerebral Mechanisms in Behavior, the Hixon Symposium*, ed. L. A. Jeffries (New York: Wiley, 1951), 1–31, on 1.

21. Years later, the biologist Sydney Brenner pointed out how von Neumann's work was much more important in his opinion than Schrödinger's, actually providing a blueprint for the mechanism of genetic inheritance: https://youtu.be/a4_JfLpZpnI.

22. John Casti, *Paradigms Lost: Images of Man in the Mirror of Science* (New York: Morrow, 1989), 132.

23. Stanley L. Miller, "Production of amino acids under possible primitive Earth conditions," *Science* 117 (1953): 528–529.

24. Jack Dunitz and Gerald Joyce, "Leslie Eleazer Orgel," *Biographical Memoirs of the National Academy of Sciences* 59 (2013): 277–289.

25. Dyson, "Comments on selected papers," 46.

26. R. Sanchez, J. Ferris, and L. E. Orgel, "Conditions for purine synthesis: Did prebiotic synthesis occur at low temperatures?," *Science* 153 (1966): 72–73; R. Lohrmann and L. E. Orgel, "Prebiotic synthesis: Phosphorylation in aqueous solution," *Science* 161 (1968): 64–66.

27. F. J. Dyson, "A model for the origin of life," *Journal of Molecular Evolution* 18 (1982): 344–350.

28. Freeman Dyson, *Origins of Life*, 2nd ed. (Cambridge: Cambridge University Press, 1999 [1985]).

29. Many years later, at the dawn of the personal computer revolution, there was an amusing exchange between Bill Gates from Microsoft and Mike Markkula from Apple about whether software or hardware is more important. Gates pointed out that hardware without software would just be a box doing nothing, drawing laughs from the audience. Markkula pointed out that software without hardware would be even more boring, drawing more laughs. See https://youtu.be/oAtTKqvfpWs.

30. Manfred Eigen, "Self-organization of matter and the evolution of biological macromolecules," *Naturwissenschaften* 58 (1971): 465–523.

31. Dyson, *Origins of Life*, 6.

32. Dyson, *Origins of Life*, 7.

33. W. Daniel Hillis, "Why physicists like models and why biologists should," *Current Biology* 3, no. 2 (1993): 79–81; see also Evelyn Fox Keller, *Making Sense of Life: Explaining Biological Development with Models, Metaphors, and Machines* (Cambridge, MA: Harvard University Press, 2002).

34. George Dyson, *Darwin among the Machines* (New York: Helix Books, 1997); see also Stefan Helmreich, *Silicon Second Nature: Culturing Artificial Life in a Digital World* (Berkeley: University of California Press, 1998).

35. M. Kimura, *The Neutral Theory of Molecular Evolution* (New York: Cambridge University Press, 1983).

36. U. Niesert, D. Harnasch, and C. Bresch, "Origin of life between Scylla and Charybdis," *Journal of Molecular Evolution* 17 (1981): 348–353.

37. See, e.g., M. Kochmanski, T. Paszkiewicz, and S. Wolski, "Curie-Weiss magnet: A simple model of phase transition," *European Journal of Physics* 34 (2013): 1555–1573.

38. Lynn Margulis, *Symbiotic Planet: A New Look at Evolution* (New York: Basic Books, 2008).

39. Lynn Sagan, "On the origin of mitosing cells," *Journal of Theoretical Biology* 14, no. 3 (1967): 225–274.

40. Michael W. Gray, "Lynn Margulis and the endosymbiont hypothesis: 50 years later," *Molecular Biology of the Cell* 28, no. 10 (2017): 1285–1287.

41. Nick Lane, *Power, Sex, Suicide: Mitochondria and the Meaning of Life* (New York: Oxford University Press, 2018).

42. One very attractive candidate for this parasitic takeover is a model by Alexander Graham Cairns-Smith at the University of Glasgow, which suggests that the early precursor to life was clay, a mineral that's abundant on the surface of the Earth. See Alexander Cairns-Smith, *Seven Clues to the Origin of Life* (New York: Canto, 1990).

43. Deborah Kelley, "Is it time to throw out 'primordial soup' theory?," All Things Considered, *National Public Radio*, 7 February 2010.

44. Kelly Kruger, Paula J. Grabowski, Arthur J. Zaug, Julie Sands, Daniel E. Gottschling, and Thomas R. Cech, "Self-splicing RNA: Autoexcision and autocyclization of the ribosomal RNA intervening sequence of tetrahymena," *Cell* 31 (1982): 147–157;

Debra L. Robertson and Gerald F. Joyce, "Selection in vitro of an RNA enzyme that specifically cleaves single-stranded DNA," *Nature* 344 (1990): 467–468.

45. Walter Gilbert, "Origin of life: The RNA world," *Nature* 319 (1986): 618.

46. Matthew W. Powner, Béatrice Gerland, and John D. Sutherland, "Synthesis of activated pyrimidine ribonucleotides under plausible prebiotic conditions," *Nature* 459 (2009): 239–242.

47. Michael Marshall, "How the first life on Earth survived its biggest threat: Water," *Nature* 588 (2020): 210–213.

48. Nick Lane, *The Vital Question* (New York: Profile Books, 2015).

49. Stéphane Leduc, *The Mechanism of Life*, trans. W. Deane Butcher (New York: Rebman, 1911).

50. M. J. Russell, A. J. Hall, and A. P. Gize, "Pyrite and the origin of life," *Nature* 344 (1990): 387.

51. Deborah S. Kelley et al., "An off-axis hydrothermal vent field near the Mid-Atlantic Ridge at 30° N," *Nature* 412 (2001): 145–149.

52. Harold J. Morowitz, Bettina Heinz, and David W. Deamer, "The chemical logic of a minimum protocell," *Origins of Life and Evolution of the Biosphere* 81 (1988): 281–287.

53. Katarzyna Adamala and Jack W. Szostak, "Nonenzymatic template-directed RNA synthesis inside model protocells," *Science* 342 (2013): 1098–1100.

54. Doron Lancet, Raphael Zidovetzki, and Omer Markovitch, "Systems protobiology: Origin of life in lipid catalytic networks," *Journal of the Royal Society Interface* 15 (2018): 20180159.

55. Daniel Segré, Dafna Ben-Eli, David W. Deamer, and Doron Lancet, "The lipid world," *Origins of Life and Evolution of the Biosphere* 31 (2001): 119–145.

56. Doron Lancet, personal communication with the author, 27 January 2021. Another scientist who has run simulations of metabolism-first scenarios is the complexity theorist Stuart Kauffman who, back in 1971, showed that the probability of having a mixture of molecules all catalyzing each other's activities (called collectively autocatalytic sets) increases with the diversity of the molecules in the mixture. See Stuart Kauffman, *At Home in the Universe* (New York: Oxford University Press, 1995).

57. Dyson, *Origins of Life*, 67.

58. H. G. Wells, *The Island of Dr. Moreau* (London: Heinemann, 1896).

59. Peter Berg and M. F. Singer, "The recombinant DNA controversy: Twenty years later," *Proceedings of the National Academy of Sciences USA* 92, no. 20 (1995): 9011–9013.

60. Sheldon Krimsky, *Genetic Alchemy: The Social History of the Recombinant DNA Controversy* (Cambridge, MA: MIT Press, 1982); Susan Wright, *Molecular Politics: Developing American and British Regulatory Policy for Genetic Engineering, 1972–1982* (Chicago: University of Chicago Press, 1994); John Durant, "'Refrain from using the alphabet': How community outreach catalyzed the life sciences at MIT," in *Becoming MIT: Moments of Decision*, ed. David Kaiser (Cambridge, MA: MIT Press, 2010), 145–163.

61. F. J. Dyson, testimony of 5 May 1977 before the Subcommittee on Science, Research, and Technology of the Committee on Science and Technology, US House of Representatives, reprinted in *Science Policy Implications of DNA Recombinant Molecule Research* (Washington, DC: Government Printing Office, 1977), 837–858.

62. Dyson, *A Many-Colored Glass*, 28–42.

63. Freeman Dyson, "Our biotech future," *New York Review of Books* (19 July 2007).

64. William H. Press and Freeman J. Dyson, "Iterated prisoner's dilemma contains strategies that dominate any evolutionary opponent," *Proceedings of the National Academy of Sciences USA* 109 (2012): 10409–10413.

65. Freeman J. Dyson, "Warm-blooded plants and freeze-dried fish," *Atlantic Monthly* 280 (November 1997): 71–80.

66. Dyson, *Disturbing the Universe*, 194–204.

67. Freeman J. Dyson, "Is life analog or digital?," *Edge.org* (13 March 2001).

68. Freeman Dyson, unpublished memo, n.d., courtesy of George Dyson.

8 THE COSMIC SEER

CALEB SCHARF

Omni magazine in 1978, interviewing Freeman Dyson: "You must be aware that some of your colleagues take a jaundiced view of your ideas about giant trees growing on comets, taking Jupiter apart to build a Dyson shell, and so on. Does it bother you to know that they're out there, muttering about 'Dyson's crazy ideas'?"

Dyson responding: "Not at all. Keep in mind, I'm also a perfectly respectable physicist, and the speculation is a hobby. It's become well known, but I've grown used to the idea that people very often become famous for accidental reasons. It's amusing to think that someday all my 'serious' work will probably be a footnote in a textbook, when everybody remembers what I did on the side! Anyway, what do I have to lose? I have tenure here, and no one expects much from a theoretical physicist once he's past fifty anyway!"

—Monte Davis, "Freeman Dyson on the Future of the Universe" (1978)

One of the most peculiar of all human traits is the degree to which we are able to think speculatively. It seems obvious that all living things speculate to some extent, whether it's a bacterium making opportunistic, noncognitive wiggles along a chemical gradient, or a raccoon eyeing the teetering lid of a well-stuffed garbage can for what could be a midnight feast. But there is little doubt that humans sit at the extreme end of this capacity, even elevating speculation to the miraculous levels of literature, art, and music.

Thinking this way has been an essential part of our species' exploration of the natural world, helping us reach for the mysteries behind the curtains. But, as they say, there is speculation and then there is speculation. Common vanilla scientific speculation can be entirely trivial, barely expending any effort to extrapolate from the body of accepted facts, or sometimes unfortunately tilting toward paper-thin fantasies that are little better than pseudoscience. Extraordinary scientific speculation, by contrast, wields deep principles and intuition for what makes physical reality tick in order to reach for nigh-on-impossible heights.

By that measure, Freeman Dyson was a mountaineer: a speculator extraordinaire, especially when it came to imagining otherworldly cosmic phenomena and life's grand trajectory through the potentially infinite future of the universe. His innate facility with such heady stuff has ensured that together with his many contributions in regular (dare I say "almost vanilla") astrophysics, his work on the side has reached far beyond the inner circles of science and into popular culture and consciousness to leave a permanent imprint.

Most of his output in these areas came from a period spanning the early 1960s into the late 1970s: a time of enormous change in the ordinary world, but also a time when scientific discovery after scientific discovery was being made in realms far beyond the confines of the Earth. In order to appreciate Dyson's most speculative work over this time, one must consider that broader context to see a little of where he was coming from.

SEARCHING FOR SIGNALS

In 1969, Dyson wrote a short *Nature* paper, "Volcano Theory of Pulsars."[1] It was only two years earlier, in 1967, that Jocelyn Bell Burnell and Anthony Hewish had reported the discovery of rapidly repeating extraterrestrial radio pulses that were subsequently understood to be spinning neutron stars (an identification strengthened after the discovery of the

Crab pulsar in 1968).[2] Although these extremely compact "nuclear balls" of matter left over from earlier supernovae explosions had been hypothesized since the 1930s, their unexpected radio emissions were what finally brought their hugely exotic nature to life for scientists and the public.

Before Bell Burnell laid the idea to rest with more data, one of the initial reactions of her colleagues to the pulsar detection was that the highly stable radio beeping might be a signal from an alien civilization, raising concern that its announcement might cause misguided public chaos.[3] We might chuckle at that scenario today, but in 1967, with the moon landings on the horizon and only a decade after the launch of *Sputnik* and the absolute zenith of the Cold War, this anecdote is a good reminder that the human zeitgeist of that period had a lot to do with unseen, looming, and hugely impactful phenomena.

Despite such compelling evidence, understanding how spinning neutron stars could launch beams of radiation at all was (and to some extent still is) a mystery. In that context, Dyson's suggestion of volcanism for an object composed entirely of nuclear matter could seem both outlandish to a layperson and perfectly reasonable to an astrophysicist. His proposal was rooted in a commonplace experience: the convective upwelling of material, like in a pot of heated water. That upwelling is good at finding a way out through any weak spots. In the case of nuclear matter, once the material reaches lower pressures at the neutron star's surface, it should expand violently, becoming a hypersonic jet—a spray of fearsome particles that might, via some mechanism that Dyson simply says, "I do not pretend to understand in detail," produce a beam of electromagnetic energy that could be detected by soft, fleshy creatures like us, scattered about the distant universe.

But perhaps more than anything else, the unexpected discovery of pulsars drove home the fact that the universe is neither silent nor unchanging. Instead it is a humming, buzzing, throbbing landscape of extreme objects and physics: a veritable dance party, if you know how to look.

A few months ahead of his paper on volcanic neutron stars in 1969, Dyson also published a study of how pulsars—depending on what was actually driving them—might be sources of gravitational waves.[4] At the time, such waves remained merely a theoretical possibility; after decades of concerted efforts, the shearing ripples in space were finally measured by the Laser Interferometer Gravitational-Wave Observatory (LIGO) detectors in Washington State and Louisiana in 2015, emanating from the distant collisions of black holes.[5] But rather than consider laboratory experiments to look for signs of these waves racing through our solar system (as the physicist Joseph Weber was doing at the time), Dyson examined the possibility that low-frequency gravitational waves, at a rumbling 1 hertz (one oscillation per second), might cause a seismic response in the Earth itself: the whole planet, or at least sections of its outer crust, serving as a gravitational wave detector. His mathematical treatment is obscenely thorough, not for the faint-of-heart, and rather sadly comes to the conclusion that the expected signal of distant pulsars might be about 100,000 times too small to be detected above the natural geophysical noise of our crunching and grinding rocky world.

At the very end of this paper, Dyson reflects on his result with a small, characteristic entreaty:

Finally, we should remember the history of radio astronomy, which was greatly hampered in its early stages by theoretical estimates predicting that few detectable sources should exist. The predictions were wrong because the majority of sources were objects unknown to optical astronomers at that time. Whenever a new channel of observation of the Universe is opened, we should expect to see something unexpected. For this reason above all, the seismic detection of pulsars is not as hopeless an enterprise as the calculations here reported would make it appear.[6]

While we've yet to identify any new phenomenon capable of strongly rattling the Earth with gravitational waves, Dyson's more general point is perfectly clear, and he himself would contribute to opening up some of those new channels of observation.

Specifically, in 1975, Dyson spent some time at the Max-Planck-Institut für Physik und Astrophysik in Munich and delivered some decidedly mainstream research dealing with the idea of adaptive optics, still quite new to most people.[7] This is a technique with which the frustrating blurring of astronomical images by the Earth's own atmosphere (the twinkling of stars so loved by everyone else) could be corrected for by manipulating and deforming a telescope's mirrors in rapid cadence. Done well, it can provide exquisitely sharp images, and Dyson, in a feat of mathematical insight, showed exactly how well it could ever be accomplished. In actuality, this work began as a classified contribution to the Jason group, seeking ways to improve satellite imagery of the Earth and telescopic missile tracking (as described in chapter 6).[8] It has become an essential astronomical technique for making the unseeable seeable.

HARNESSING THE STARS

Throughout this same period, Dyson clearly also had his mind on areas lurking even farther over the horizon. In 1960 he published a one-page paper in the prestigious journal *Science* that contained the seeds for what would become one of the most widely recognized named hypotheses outside the academic world: the Dyson sphere.[9]

At the core of his idea was a pretty basic question about what happens if an energy-hungry civilization grows at a consistent percentage rate of, say, 1 percent total per year. The answer is that after three thousand years, there will have been growth by a factor of over 10^{12}, or a *trillion* times. Such growth firmly exceeds the Malthusian limit posited by Thomas Malthus in 1798 as the point by which a human population would run out of food, water, and space.[10] But what Dyson was really chasing was the question of whether there would be external evidence of such an overrun species, the kind of evidence that a keen terrestrial astronomer might be able to see using newly sensitive infrared detection technology.[11] In other words, Dyson, who refers to then quite novel efforts in radio SETI (the Search for Extraterrestrial Intelligence), saw an exciting engineering advance in infrared astronomy and went in search of something really interesting for it to look at. It also marks a topic that was of enormous interest to him: the question of other technological species out there in the cosmos and our own human future.

In his short paper, he suggests a trajectory that a fit-to-burst species might follow in search of its lebensraum (or "living space"). Naturally, he posits, such a species would want to make the most efficient possible use of the greatest energy source available: its star. Equally naturally, according to Dyson, the species would end up needing to disassemble a large planet's worth of matter in order to construct the necessary collectors of stellar photons. What is so delightfully mad (or at first seems that way) is that to compute the feasibility of this effort, he dispatches a "planet such as Jupiter" by estimating the energy required to tear apart and repurpose its matter. The answer: a mere eight hundred years worth of the total output of the sun. The end result would be some kind of structure completely surrounding the star at a distance equivalent to that of the Earth from the sun. The habitable space on its interior surfaces would total an area that could be 550 million times that of Earth's surface.

But the most critical feature for Dyson was that this orbital monster would have to obey the laws of thermodynamics. Consequently, whatever power it absorbed and used from the star would have to be reemitted by its dark outer side as thermal radiation, or waste heat. With a total power output equal to that of a sun-like star but entirely in the infrared portion of the electromagnetic spectrum, such a technological structure would present a special signature for infrared astronomers.

Dyson was the first to admit that his conception was not original. He had read the classic science-fiction tale *Star Maker*, written by the British philosopher and author Olaf Stapledon in 1937, in which the history of life in the universe (and others) is imagined, together with gauzy "light traps" surrounding entire star systems.[12] But what Dyson brought to the table was an unerring quantitative argument and a potentially testable hypothesis. Like any other good hypothesis, it was also quite capable of being disproven. As infrared astronomy became more sensitive, scores of previously unseen star-sized sources of radiation were discovered dotted across space. These weren't Dyson spheres, though; they were protostars, the baby gatherings of nebular matter not yet hot enough to trigger stellar fusion.

The starship *Enterprise* near a Dyson sphere, as featured in the episode "Relics" from the popular television show, *Star Trek: The Next Generation*, originally broadcast on October 12, 1992. Source: Getty Images.

When I first read Dyson's 1960 paper, I was astonished at how convincingly he could pull off such an outrageous-sounding sequence of proposals. He doesn't actually say very much at all in his short paper. Instead the reader is carried along by the stark clarity of his physics and, to be perfectly honest, the calm conviction that this is all perfectly sensible. (I have often wondered whether Douglas Adams knew about Dyson's ideas when he came up with the Magratheans—the designers and builders of bespoke planets.[13] My best guess is absolutely definitely yes, but also maybe not.)

Several years later, in 1966, Dyson put into print a much more richly developed set of arguments surrounding the notion of restructuring a planetary system in service of energy capture. This came in the form of an article, "The Search for Extraterrestrial Technology," published in a conference volume in honor of the sixtieth birthday of his former doctoral adviser, physicist Hans Bethe.[14]

In this longer foray, Dyson elaborates on his technical proposals and his motivations. His feeling is that while we might desperately want to engage with "intelligence" elsewhere, what we can actually do is look for the remote signs of technology (a viewpoint that is now, in the 2020s, being once again put forward vigorously by astronomers and astrobiologists). Furthermore, whereas we cannot know what a "very large and conspicuous technology would be interested in doing," we can definitely consider what it *has to do* to exist at all.

It's here that Dyson clarifies that an energy-capturing structure around a star has to be in discrete pieces. Put simply, a single mechanically interconnected sphere surrounding a star (or any mass) will be inherently unstable to the star's gravitational forces and likely to fall off balance and onto the star itself. The proper approach is to build many orbital pieces—in effect, a swarm—at staggered distances and distributed so as to capture most photons as they stream from the star but not as a monolithic cloak.

He computes the stresses and material qualities and the plausible geometries of the building blocks for this sieve-like structure (octahedra seem promising). But where does the matter come from to make something so gargantuan? In 1960 he used the disassembly of Jupiter as an example without providing the mechanics of that process. In his 1966 analysis, he puts forth a physics-driven engineering proposal for disassembling a world like the Earth. He indicates, with typical dry humor, that he is not suggesting such planetary disassembly be put into practice; he simply uses the example to compute the magnitude of the challenge.

The approach he suggests is to bind the surface of the planet with conductive wires along lines of latitude and then, via an orbital structure using solar energy, to drive the spin-up of the planet with a magnetic field, as if it were a giant brushless electric motor. (The details he provides are in truth rather more complex, handling the transfer of angular momentum from the orbital structure into the solid-body rotation of the planet, but this is the general principle.) Eventually the planet can be spun up so fast that it literally starts to come apart at the seams and conveniently feeds itself into orbiting pieces ready to be whisked off to build a starlight-capturing megastructure. The total power requirements? About three hundred times the solar power received by the Earth in its current state—and hence eminently plausible.

Encouraged by this hypothetical success, Dyson goes even further to discuss how to disassemble stars themselves so that their matter can be deployed across interstellar distances to do things like build extra planets (needed as bases for efficient galactic exploration). One solution he suggests would be to engineer star-on-star collisions in order to strip their matter. A consequence of this astroengineering would be that the character of entire galaxies might wind up being divided between "wild" types and "tame" types. An extensively engineered tame galaxy, for example, might show distinct excesses of blue and infrared light from all of its technological repurposing.

All of these projections about what might happen for a technological species as its footprint in the cosmos grows were deeply interesting and appealing to Dyson—even if, somewhat ironically, his calculations suggested that such extreme "makeovers" of astrophysical environments cannot be rife in our own Milky Way, or else we might have noticed. In this 1966 paper and in others, he reiterates the general motivation that our own growth may eventually make engineering on an astronomical scale necessary. Therefore, thinking about what kind of signatures very advanced forms would present to us from afar is also a way to probe our own future.

In between these published works on the Dyson sphere concept came another wonderfully extreme idea: the gravitational machine. In a 1962 essay in *Interstellar Communication*, he investigates the prospects for somehow harnessing the energy of gravitational fields.[15] These are pitifully weak on small scales yet monstrously potent on large scales. His proposal starts off relatively innocently with a version of the so-called slingshot mechanism widely used by interplanetary spacecraft: during a close pass to an orbiting planet, a vehicle can "steal" a tiny bit of that planet's momentum by allowing itself to be pulled along by the planet's gravity for a short while. Because of the overall conservation of momentum, the spacecraft can gain significant velocity, while the planet loses only a teeny bit because it is so much more massive.[16]

In Dyson's machine, a pair of stars orbit each other, while a planet orbits farther out and around this pair. Small objects like spacecraft can be "dropped" from the planet in such a way that they will loop around one of the orbiting stars and get a slingshot boost—stealing a little of the momentum and gravitational energy of the binary star system. With ordinary stars like the sun, this is a useful, albeit rather low-output, way to obtain energy, since there is much more energy available from the radiation output of the stars. But if the stars were more physically compact white dwarfs (the remains of stars that have completed active fusion processes), the situation could be reversed. Orbiting much closer together—perhaps with periods of only 100 seconds—such speedily moving stellar remnants could readily boost an object to velocities of 2,000 kilometers per second (about 0.7 percent of the speed of light).

A critical feature of this kind of boost is that because the spacecraft or object is freely falling in the star's gravitational field, the internally felt forces would be zero, except for mild stresses due to tidal effects, even though the acceleration as measured in an external reference frame could reach an equivalent of 10,000 g—a rather neat consequence of Einstein's equivalence principle, whereby gravitation and acceleration are indistinguishable phenomena. This feels like a characteristic piece of Dyson's genius: a fundamental physical insight that yields something quite miraculous when applied in the right place. He goes on to suggest that a species might engineer these white dwarf pairs, picking their configuration to yield the desired results and placing them around the galaxy to serve as handy boost points: the Tesla superchargers of the cosmic jet set and potentially detectable as anomalies by Earth-based astronomers.

Naturally this wasn't the end of his speculations. The universe produces far more compact and extreme collections of mass than white dwarfs, and in 1962 neutron stars were still an intriguing theoretical proposition (before their detection as pulsars). Dyson muses on the possibility of neutron star pair boosters. But he realizes that whereas these much more compact remnants (a mere 10 kilometers across compared to a white dwarf that is a little larger than the Earth) could be set up to offer greater velocity boosts (up to some 27 percent of the speed of light), they would also vigorously radiate away their own orbital energy as gravitational waves. That energy loss would be rapid enough to make neutron star pairs quite an unstable configuration. They would spiral together too quickly to be of more than limited practical use; such a rapid inspiral, Dyson continues, would send out ripples in space-time detectable by a suitable instrument hundreds of millions of light years away. And so it was that Dyson foresaw the eventual detection (via gravitational waves) of binary neutron-star collisions nearly fifty years before the LIGO and Virgo collaborations succeeded in detecting just such an event—albeit, we presume, that the 2017 detection stemmed from the collision of naturally occurring neutron stars rather than the collapse of an engineered gravity-boost device.[17] Dyson's insights provide a potent example of how extraordinary speculation—in the right hands—can foreshadow conventional (if still spectacular) science.

THE LONG VIEW

Examining all of this today, it's difficult not to feel that throughout these astrophysical works, Dyson was testing the waters for something more. He seems to have been dipping his toes into a pool of ideas, finding intriguing avenues and useful connections between bits of physics, bits of engineering, and bits of futures that could be. Whether this was really the case (above all, I suspect he was just having fun), eventually this rich mixture coalesced in one place, a surprising and profound nexus: an encoding of thought in one special restructuring of matter in paper and ink, electrons and bits.

"Time without End," published in July 1979 in the *Reviews of Modern Physics*, is perhaps Dyson's most astonishing and challenging paper of this period.[18] Its subtitle is "Physics and Biology in an Open Universe." This lengthy article stemmed from four lectures that Dyson first delivered at New York University in 1978. In it he surveys the history and philosophy of cosmology, evaluates the characteristics of an endlessly expanding cosmos, and constructs a series of mathematical physics-based arguments for how life could continue effectively forever. Even more: he pursues what would be required for communications between life in such a universe to continue effectively forever and for the potential of connecting all talkative species to all others for eternity. It is best read sitting down with a stiff drink to hand.

Not only are his arguments pithy and clever; as always, he builds them from scratch from fundamentals, incorporating tools like a quantum mechanically based description of "subjective time," information theory, and information entropy. His arguments further draw on the raw qualities of electrons, the emission and absorption of electromagnetic radiation, and the cosmic-scaled geometry and evolution of space-time itself.

Dyson began his 1978 lectures at the end. He pointed out that whereas scientific attention had been lavished on the Big Bang, little had been written about the future of the universe, or more specifically, eschatology, the study of the end of the universe. He argued that one simply cannot study the long-term future of the cosmos without considering the role of intelligence and its impact on that future. In a universe that stops expanding and recontracts, the future is one in which Dyson says "we have no escape from frying," although there might still be a trillion years during which one could try to find a technological fix. By contrast, he proposes that an open, endlessly expanding universe offers much more hopeful, although still challenging, possibilities.

The launching point of his technical discussion is to survey all the things that will happen to the universe as time passes. That includes the evolution of stars into more and more quiescent objects, from white dwarfs and neutron stars to black holes, all during the next few trillion years. Furthermore, he posits, planets will suffer from orbital disruption and become scattered as stars randomly encounter other stars (each laden with planetary children) over a few hundreds of trillions of years within a galaxy. In turn there will be a dynamical cooling of galaxies as stars themselves are scattered outward, effectively evaporating like molecules of gas (over 10^{19} years). Even the end points of matter represented by black holes will evaporate because of Hawking radiation (on timescales up to 10^{100} years). Also, if condensed matter, like rocks and planets, cools toward absolute zero, it will, over timescales of 10^{65} years, behave like a liquid because of quantum-mechanical tunneling. When we consider other effects near absolute zero on timescales of 10^{1500} years, tunneling-driven nuclear reactions in ordinary elements will eventually cause all ordinary matter to fuse into iron nuclei. The bothersome question then is whether all stars will eventually collapse into just black holes (over spans of 10-to-the-10^{26} years) or into neutron stars and black holes (10-to-the-10^{76} years). The answer depends on assumptions about as-yet-unknown microphysics.

It is quite a collection of gloomy-sounding phenomena. But the real purpose of this foray into what happens over such frighteningly long timescales is to demonstrate that stuff *does* happen even over such frighteningly long timescales. And that is an essential ingredient for bringing life into the equation.

To do this, Dyson discusses the idea of embodiment—namely, that (if I translate into the kind of language we might use today) life need not be constrained to one substrate; it could easily be "postbiological" in nature. Instead, what is most important is whether life can find a way to continue itself, to produce copies in any way possible. To capture this phenomenon Dyson proposes a universal scaling law, based on the energy involved in copying one quantum state to another quantum state, together with the environment of that state (defined by its temperature). This, he indicates, is surely the most fundamental statement about what life does, and in its quantum-mechanical formulation, he arrives at its most irreducible description.

Next he packs two jaw-dropping consequences into the space of a few paragraphs: first, that "subjective time, " rather than physical time, is the correct measure of time for a living entity, and second, that any living entity is characterized by its rate of entropy production per unit subjective time, a quantity he labels "Q."[19] In other words, whereas physical time in the universe might tick away in the metronome-like frequencies of electromagnetic radiation or the decay of unstable isotopes, for life the clock runs by changes in configuration, by *processes*: one entropy-producing chemical reaction after another, one thought after another, in an entirely subjective way.

Even more explicitly, the use of entropy lets him connect life to information theory and to the fundamental relationship between entropy and information, so that Q for a living thing is also the amount of information (in bits) that must be processed to maintain aliveness—or, as Dyson himself puts it, "in order to keep the creature alive long enough to say 'Cogito, ergo sum.'"[20] In other words, Q is actually a measure of the complexity of life. By this standard, he estimates that a human must have a Q of about 10^{23} bits, and all of humanity a Q of 10^{33} bits.

This processing of bits, though, the flow of Q, has to involve the dissipation of energy, and that dissipation has an intimate connection to the temperature of the environment.[21] If the temperature of the environment is equal to or greater than the temperature of life, then there can be no dissipation and no processing. In other words, to maintain itself through Q, life has to have a heat sink. The ultimate heat sink is the universe itself (with a present temperature of around 3 Kelvin), but in a future of steady expansion and eventual loss of things like warm planets and hot stars, there is a danger of life reaching a thermal equilibrium with the cosmos that would prevent it from sustaining its necessary Q. To set the stage for dealing with this, Dyson yet again drills into the most fundamental physics for guidance. This time it's the physics of energy loss and dissipation through electromagnetic radiation.

Most people I know, myself included, would just try to lift a formula from a textbook and cobble an answer together, but Dyson needs to know the absolute upper limit to this dissipation and admits that he can't find it in textbooks. Instead, he provides a derivation from first principles of a formula that invokes the details of radiating electromagnetic dipoles for oscillating electrons and some hairy matrix mathematics. The upshot is that, on the face of it, with a finite store of energy for life (as is the case on Earth with our finite sun, or even for a civilization with a whole tame galaxy at its disposal), no matter how efficient such a civilization might be, it will eventually run out of ways to sustain itself. Even slowing one's metabolism is not going to help. It is at this point in the paper that I usually need to go for a soothing walk to remind myself that Dyson is about to reveal a firecracker of an idea, an escape clause, a way for life to continue forever in a universe of ever dwindling action.

To try to put his fantastic proposal into simple terms: He rallies his mathematical insights to present a solution that resembles Zeno's famous paradox, the eternal crawl toward a finish line where each step is a fraction of the remaining distance. In Dyson's case, that crawl allows life to continue to function, with consciousness, for an unlimited amount of time in an expanding universe, despite having a finite amount of energy at hand. The trick to this solution lies in the nature of subjective time and in hibernation.

The first part of the trick relies on the fact that if you can wait long enough for the universe to cool as it expands, the energy that was useless in your immediate environment during one epoch becomes useful again. To put this another way, once your temperature matches the temperature of the universe around you, there is nothing to do but wait until the universe has gotten cooler so that you can once again exploit that temperature difference to do stuff, like metabolize, think, and run your machines. The second part of Dyson's trick is that as time goes by, life can adjust its "duty cycle" by hibernating for longer and longer periods, interspersed with moments of activity. By tuning, or lowering this duty cycle in proportion to temperature as physical time elapses and with subjective time being uncaring of the physical time, Dyson argues that life can happily continue literally forever, without an infinite energy supply.

To put this slightly differently, as the universe ages, you may have to wait longer and longer between smaller and smaller sips of your finite supply of energy—in order for it to be thermodynamically possible to use that

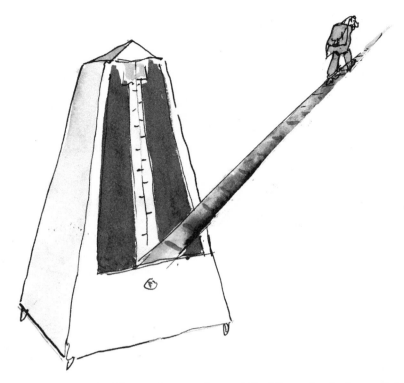

"As time goes by, life can adjust its 'duty cycle' by hibernating for longer and longer periods."

energy—but from your subjective perspective, time is flowing along normally. The universe around you may seem to flicker toward a cooler and more expanded state like a glitchy movie, but you can keep reading that enjoyable novel or thinking that intriguing thought in perfect continuity.

Of course, there is some sleight of hand at play here, as Dyson readily admits. For example, can metabolism actually operate at temperatures close to absolute zero as the universe continues to cool? And there is the thorny question of memory capacity in infinite subjective time, or as he puts it with a certain wry humor:

I would like our descendants to be endowed not only with an infinitely long subjective lifetime but also with a memory of endlessly growing capacity. To be immortal with a finite memory is highly unsatisfactory; it seems hardly worthwhile to be immortal if one must ultimately erase all trace of one's origins in order to make room for new experience.[22]

This raises the specter of an unavoidable limitation for digital storage because of the finite number of atoms ever available. Instead, he muses, perhaps an analog system—such as using the angle between objects like stars in the sky—could offer the infinite capacity needed. This was a proposal that he came back to in later years as others refuted the idea of infinite life in an evolving universe as well as the notion of analog life and analog memory.[23]

Finally, Dyson deploys a similar, but slightly more intricate, argument to the one for infinite life, this time to address whether communication between beings in distant galaxies within his hyperbolically expanding universe could continue indefinitely. Once again, he dives into the core cosmological physics and equations of electromagnetism, together with information theory (handling bit rates and, critically, noise from the cosmic background radiation), to render a proposal for communication comparable to his arguments for life's infinite subjective longevity. Within a range of distances and with evolving duty cycles, two civilizations should be able to continue passing information between each other forever, because there is an ever-declining energy requirement for transmitting a fixed amount of information. In other words, a finite energy supply suffices to let you send 280-character texts for all time.[24]

There are limits, though. Societies too far apart would not be able to pull this trick off, at least for direct communication, although with strategic use of relay stations, in principle every talkative group in the universe could eventually be connected to every other one. It is this last point that seems to capture one of Dyson's own core qualities: his boundless capacity for scientific and philosophical optimism. Indeed, at the very end of "Time without End," he points out a difference of inclination between the physicist Steven Weinberg and himself. Weinberg, a master of catchy but profound statements, had recently written that "the more the universe seems comprehensible, the more it also seems pointless."[25] Dyson seems quite delighted to be able to push back on this nihilistic take on things with his demonstration of the hypothetical capacity of the universe to not only grow in complexity and richness without limits but to also become ever more interconnected, with its many minds forever in conversation.

Dyson wrote "Time without End" about a decade after the hot Big Bang had become well established within the scientific community; for example, the cosmic microwave background radiation, widely understood to be the remnant glow from the Big Bang era, had been detected in the mid-1960s. But during the 1970s, when Dyson pursued his cosmological work, neither the shape of space-time on cosmic scales nor the dynamics by which it evolved were well understood.[26] At the time, astronomical data seemed most consistent with the idea of an expanding, "open" (hyperbolic) universe, in which the average density of matter throughout space was less than (although still close to) the so-called critical density—the density that would make the universe forever slow in its expansion but never recollapse, a situation of perfect balance. In the early 1980s, a new paradigm emerged around the idea of cosmic inflation, whereby the youthful universe would have experienced a brief period of astonishingly aggressive exponential growth. Inflation helped to address a number of nagging issues with the standard Big Bang account, including the remarkable isotropy and homogeneity of the observable universe (its statistical uniformity in all directions), as well as why the universe should be so close to that critical density—why, in other words, the shape of space across the longest distances seemed to obey the familiar Euclidean geometry of schoolroom mathematics, where straight lines are indeed straight lines.

By the late 1990s, a new puzzle emerged as measurements of the rate of cosmic expansion over the past few billions of years suggested that somehow the present-day cosmos was accelerating in its expansion: not only getting bigger, but getting bigger at a faster and faster rate. This discovery, now bundled into the hypothesis of "dark energy," does indeed indicate that the universe should keep expanding forever. But it presents an additional challenge for Dyson's proposals because in an accelerating cosmos, any structures of matter (such as galaxies) will eventually become utterly isolated from each other, with any meaningful contact effectively outstripped by the expansion of space.[27] Even more dramatic, if the acceleration of expansion continues, there may come a time when the forces that bind structures together—from gravity to electromagnetism—can't compete and everything disintegrates into finer constituents.

On the face of it, these later cosmological developments render some of Dyson's proposals problematic, which Dyson himself readily admitted as pieces of the new cosmological picture emerged. Nonetheless, as Dyson himself had mused in a different context, the possibility that the world may be different than we think is not sufficient reason to avoid asking our questions. It is difficult not to want the universe that Dyson found to be so gloriously full of future promise. And the simple truth is that we may still find that we are in such a reality as new cosmological discoveries are made. In the end, he might get the last, optimistic laugh.

STAR CHILD

Dyson's boundless optimism has surely contributed to the enduring popularity of so many of his speculations, boosted by his talent for writing many accessible versions in books like *Disturbing the Universe, Infinite in All Directions,* or *Imagined Worlds.*[28] The few examples of Dyson's astrophysically inclined work described here make it easy to see why his ideas have provided such fodder for others hoping for inspiration and cosmic pizzazz. That's especially true for our enduring questions about intelligent, technological species elsewhere in the universe—a subject that Dyson saw more clearly than most others to be a genuine scientific puzzle that could be addressed with the same tools we might apply to any other natural phenomenon.

Search for "Dyson spheres" on Google and you'll find over 4 million results. On Wikipedia, an entire page is devoted to "Dyson spheres in popular culture," with references ranging from dozens of novels to comics, television shows, and movies, including *Star Trek, Futurama,* and *Avengers: Infinity War,* in which the world Nidavellir is depicted as a Dyson structure surrounding a neutron star, because, well, why not.[29] There are also, perhaps unsurprisingly, an astonishing number of video games that use the general concept. There is no doubt that Dyson has been a muse for other creative minds in the twentieth and now twenty-first centuries and will likely be one for future centuries.

Back in the late 1960s, when filmmaker Stanley Kubrick was conducting research for his iconic film *2001: A Space Odyssey,* he talked with Dyson, and one can't help but imagine what a fantastic conversation that must have been.[30] It would have come sandwiched midway between Dyson's proposals of sun-enveloping structures and the astonishing concept of never-ending life. In retrospect, Dyson was the original "star child," an intellect whose own odyssey reached far beyond its Earthly manifestation.

NOTES

1. F. J. Dyson, "Volcano theory of pulsars," *Nature* 223 (1969): 486–487.

2. A. Hewish, S. J. Bell, J. D. H. Pilkington, P. F. Scott, and R. A. Collins, "Observation of a rapidly pulsating radio source," *Nature* 217 (1968): 709–713; David H. Staelin and Edward C. Reifenstein III, "Pulsating radio sources near the Crab Nebula," *Science* 162 (27 December 1968): 1481–1483. The radio telescope used by Bell Burnell and Hewish seems outlandishly crude by modern standards. It consisted of a thousand posts planted in about four and a half acres of a Cambridgeshire field (roughly fifty-seven tennis courts' worth of area), strung together with 2,000 dipole antennae made from about 120 miles of wire and cable.

3. George Hobbs, Dick Manchester, and Simon Johnston, "Fifty years ago Jocelyn Bell discovered pulsars and changed our view of the universe," *Conversation* (27 November 2018), https://theconversation.com/fifty-years-ago-jocelyn-bell-discovered-pulsars-and-changed-our-view-of-the-universe-88083.

4. F. J. Dyson, "Seismic response of the Earth to a gravitational wave in the 1-Hz band," *Astrophysical Journal* 156 (1969): 529–540.

5. B. P. Abbott et al. (LIGO Scientific Collaboration and Virgo Collaboration), "Observation of gravitational waves from a binary black hole merger," *Physical*

Review Letters 116 (2016): 061102. On the long history of searching for gravitational waves, see Harry Collins, *Gravity's Shadow: The Search for Gravitational Waves* (Chicago: University of Chicago Press, 2004); Daniel Kennefick, *Traveling at the Speed of Thought: Einstein and the Quest for Gravitational Waves* (Princeton: Princeton University Press, 2007); and Janna Levin, *Black Hole Blues and Other Songs from Outer Space* (New York: Knopf, 2016).

6. Dyson, "Seismic response," 540.

7. Freeman J. Dyson, "Photon noise and atmospheric noise in active optical systems," *Journal of the Optical Society of America* 65 (1975): 551–558.

8. Freeman J. Dyson, "Comments on selected papers," in Dyson, *Selected Papers with Commentary* (Providence, RI: American Mathematical Society, 1996), 1–49, on 42. On the Jason group, see Ann Finkbeiner, "Jason—a secretive group of Cold War science advisers—is fighting to survive in the 21st century," *Science* (27 June 2019), available at https://www.sciencemag.org/news/2019/06/jason-secretive -group-cold-war-science-advisers-fighting-survive-21st-century. Dyson's work fits a larger pattern at the time of cross-overs between classified projects and advances in basic physics and astronomy. See, for example, Benjamin Wilson and David Kaiser, "Calculating times: Radar, ballistic missiles, and Einstein's relativity," in *Science and Technology in the Global Cold War*, ed. Naomi Oreskes and John Krige (Cambridge, MA: MIT Press, 2014), 273–316; Benjamin Wilson, "The consultants: Nonlinear optics and the social world of Cold War science," *Historical Studies in the Natural Sciences* 45 (2015): 758–804; and Neil DeGrasse Tyson and Avis Lang, *Accessory to War: The Unspoken Alliance between Astrophysics and the Military* (New York: Norton, 2018).

9. F. J. Dyson, "Search for artificial stellar sources of infrared radiation," *Science* 131 (1960): 1667–1668.

10. Malthus's essay was originally published without attribution: *Essay on the Principle of Population as It Affects the Future Improvement of Society, with Remarks on the Speculations of Mr. Goodwin, M. Condorcet and Other Writers* (London: J. Johnson in St. Paul's Church-yard, 1798).

11. Helen J. Walker, "A brief history of infrared astronomy," *Astronomy and Geophysics* 41 (October 2000): 10–13, on 10.

12. Olaf Stapledon, *Star Maker* (London: Methuen, 1937).

13. Douglas Adams, *The Hitchhiker's Guide to the Galaxy* (London: Pan Books, 1979).

14. F. J. Dyson, "The search for extraterrestrial technology," in *Perspectives in Modern Physics: Essays in Honor of Hans A. Bethe*, ed. R. E. Marshak (New York: Wiley, 1966), 641–655.

15. Dyson's essay was "Gravitational machines," in *Interstellar Communication*, ed. A. G. W. Cameron (New York: W. A. Benjamin, 1963), 115–120.

16. This approach, sometimes also called a "gravity assist," uses the orbital motion of planets to provide velocity boosts. Most famously it enabled the "grand tour" of the *Voyager* spacecraft and continues to be employed for many interplanetary missions to alter velocities and trajectories.

17. B. P. Abbott et al. (LIGO Scientific Collaboration and Virgo Collaboration), "GW170817: Observation of gravitational waves from a binary neutron star inspiral," *Physical Review Letters* 119 (2017): 161101.

18. F. J. Dyson, "Time without end: Physics and biology in an open universe," *Reviews of Modern Physics* 51 (1979): 447–460.

19. I cannot resist pointing out that in the long-running *Star Trek: The Next Generation* television series (1987–1994), an important recurring character is an omnipotent being called simply "Q."

20. Dyson, "Time without end," 454.

21. The dissipation and flow of energy was a subject that Dyson discussed in several places, including a lengthy popular-science piece written for *Scientific American*: Dyson, "Energy in the universe," *Scientific American* 225 (1971): 50–59.

22. Dyson, "Time without end," 456.

23. Dyson discusses this in a piece written in 2001 for *Edge* magazine and in his responses to an online discussion, accessed 29 December 2020 at https://www.edge.org/conversation/freeman_dyson-is-life-analog-or-digital.

24. That prospect of an eternity of social-media-style communication might make you wince, but the real point is that 280 characters represents a fixed (or close to fixed) amount of information.

25. This statement comes at the end of his book: Steven Weinberg, *The First Three Minutes: A Modern View of the Origin of the Universe* (New York: Basic Books, 1977), 154.

26. This is a lengthy story, involving both the detection of the cosmic microwave background by Arno Penzias and Robert Wilson in 1964, as well as the theoretical

predictions and work by several others. An excellent first-person history is given by Nobel laureate Jim Peebles in his recent book: P. J. E. Peebles, *Cosmology's Century: An Inside History of Our Modern Understanding of the Universe* (Princeton, NJ: Princeton University Press, 2020). See also Helge Kragh, *Cosmology and Controversy: The Historical Development of Two Theories of the Universe* (Princeton: Princeton University Press, 1996).

27. L. M. Krauss and R. J. Scherrer, "The return of a static universe and the end of cosmology," *General Relativity and Gravitation* 39 (2007): 1545–1550, arXiv:0704.0221.

28. Freeman Dyson, *Disturbing the Universe* (New York: Harper and Row, 1979), *Infinite in All Directions* (New York: Harper and Row, 1988), and *Imagined Worlds* (Cambridge, MA: Harvard University Press, 1997).

29. See https://en.wikipedia.org/wiki/Dyson_spheres_in_popular_culture (accessed 29 December 2020).

30. Phillip F. Schewe, *Maverick Genius: The Pioneering Odyssey of Freeman Dyson* (New York: Thomas Dunne Books, 2013), 149–151.

9 A BOUQUET OF DYSON

JEREMY BERNSTEIN

The first time I met Freeman Dyson—at least in a manner of speaking—
was in fall 1953. He had come to Harvard to give a series in the Morris Loeb
and David M. Lee Lectures in Physics. Some of them were on a subject on
which I was then doing my PhD thesis—the theory of mesons. My thesis adviser, the late Abraham Klein, suggested that we have a private talk
with Dyson. I was very pleased with this idea because Dyson was already
a hero of mine for his monumental work on quantum electrodynamics.
His first lecture at Harvard had been about something else. He had been
introduced by our local genius Julian Schwinger as "mister" Dyson, which
I took to be one of those professorial affectations. But I later learned that it
was true. While he has been awarded honorary degrees from over twenty
institutions, Dyson never did bother to get his PhD.

He had been assigned a temporary office in the physics building, which
we had no trouble finding. The office had a couch on which Dyson was
lying, seemingly asleep. This did not bother Klein, who began to drone
on and on, oblivious to the fact that Dyson was lying there with his eyes
closed. After some time, it registered on Klein that there was no response,
so we left. I did not see Dyson for another four years, by which time I had

gone as a visiting member to the Institute for Advanced Study, at which he was a professor. He occasionally had lunch with us at the Institute cafeteria. He rarely said much but seemed to be amused by the general patter. I never would have gotten to know him except by accident: we had taken the same late-night train from New York to Princeton, so we began to talk. As I recall, I had my car at the station, and I drove him home. He invited me in for a drink and wanted to know how, and why, I had gotten into physics.

Toute proportion gardée, our routes had been similar: mathematics. There was an important difference: he was a genius. It may have been on that occasion that I asked him what his earliest mathematical memories were. He said that when he was still young enough for being "put down for naps," he began adding up $1 + 1/2 + 1/4 + 1/8 + 1/16 \ldots$ and realized that the sum was approaching 2. In short he had grasped the idea of a convergent infinite series. At the time of our meeting, he was separated from his wife. His children, George and Esther, were for the moment with her, so he was living in this big house alone. We talked about that.

During the year, I saw him to talk a few times. He was clearly very busy. One of the things he did was to read and gloss over papers in Russian journals that were not yet translated. Once when I went into his office, he was relaxing in a canvas beach chair and reading the Bible in Russian. That summer he went off to the General Atomic division of the large firm General Dynamics in La Jolla to consult, and I went to the RAND Corporation in Santa Monica to do likewise. It was not a success.

Although RAND had the superficial atmosphere of a college campus, it was devoted to the strategy of nuclear war. The Falstaffian figure of Herman Kahn was assuring everyone that a few megadeaths in an exchange with the Soviets would be quite tolerable. I recall once going into a room with the other RAND physicists where a seismograph had been set up. We watched while it registered the quavers from a hydrogen bomb test in the Pacific. I found the whole atmosphere very depressing. In the meanwhile, the secretary we had in our building at the Institute was forwarding mail along with the gossip. I learned that Dyson was designing a spaceship, that he had been to a bullfight, and had been bitten by a dog. I wrote a note to him saying that if any of these three things were true, he was having a better time than I was. Much to my surprise, a day or so later the phone rang, and it was Dyson inviting me to come down to La Jolla. I jumped at the chance.

It turned out that all three of these things were true, and the spaceship, which was supposed to be powered by the exploding atomic bombs, was called the "Orion." It is not an accident that this was the name of a spaceship in Kubrick's 2001. There was more. He had been stopped by the police for walking. They had some reason: he had broken his glasses and was wearing scuba diving goggles to assist his vision. When asked for identification, he produced a card with his picture and fingerprints on it from the Department of Defense. It said that the bearer of this card was entitled to receive top-secret information. One can only wonder what went through the police officer's mind.

One of the considerations in designing the Orion was how much radiation would propagate through the superheated material following an explosion—the opacity. Dyson had proven a beautiful theorem that showed that quantum mechanics imposed a maximum. My job was to test this with numerical computations, which I did, on a now antique electromechanical calculator. Some fifty years later, we published this work.[1] In the meantime, a great deal had happened in his life, much of which began that summer.

When I arrived in La Jolla, the children were still with their mother. She and Freeman were divorcing, and the children were to come and live with him. When they arrived, they were accompanied by an attractive German woman named Imme Jung, who was to act as their governess, allowing Freeman the opportunity to go to work. He was totally preoccupied with the spaceship and wrote a stream of reports on everything from the functioning of springs to how to shape the explosive force from an atomic bomb. As the summer evolved, two complications developed. As Imme was scheduled to return to Germany, a second German governess named Margot had arrived as her replacement. But in the meanwhile, Freeman and Imme had fallen in love, their happiness evident to everyone. The four of us took to going out together, on one occasion driving to Baja California and on another to visit some people I knew who had a ranch near Santa Barbara. By the end of the summer, Freeman and Imme had decided to get married. He was going to stay in La Jolla, and I was going to return to Princeton.

We began to write letters. I saved his, all written in an almost calligraphic script. The first one is dated 25 November 1958. And the last one I have is dated 26 December 1990. We then stopped communicating by letter and email took over. I have dozens of emails from Freeman. Some of them are very interesting, but they lack the human touch of an actual letter. Can you imagine Lord Chesterfield's "Tweets to His Son"? Some of the letters are technical. One of them, dated 21 August 1981, begins, "Sorry my friend you goofed." I had written a book review for the *New Yorker* that had dealt with maps and had carelessly described the Mercator projection—which projects the spherical Earth onto the plane of a map—as "linear."[2] Dyson writes, "Mercator's is not a linear projection but a logarithmic one. So far as I know there is no geometric model for it. Precisely that was the originality of Mercator's idea." He then goes on to provide a mathematical proof, and what is typical, he goes on to say, "The really extraordinary thing about Mercator's projection is that it gives useful maps up to very high latitudes. This is possible only because the latitude scale is logarithmic." He shows it works well, up to a latitude of 85 degrees, and remarks that the "map includes more than 99% of the globe. No geometric projection could possibly do as well as this," and then adds, "It is a pity that schools do not teach geography anymore."

He was of course right, and my error also slipped by the *New Yorker*'s vaunted checkers. The interested reader can find the details on the web.

Many of the letters are anecdotal. Typical is the first one dated 28 November 1958. It was written just after Freeman and Imme had gotten married. It was addressed to "Jerry and Jane." Until I began writing for the *New Yorker*, I was always known as "Jerry" instead of my real name "Jeremy." "Jane" refers to Jane Kane, who was the secretary who had kept me informed about Freeman's activities while I was at RAND:

Dear Jerry and Jane,

Your telegram made us very happy when it arrived on the wedding morning.

The wedding was a merry occasion. The judge took us in a break between two traffic violation cases and the children found this highly entertaining. We sat in the courtroom with the flowers and the 3 children [two of his own and one of his ex-wife's] and our witnesses . . . while the lawyers cross-examined some poor fellow who had been crashed into by a car-load of drunk Mexicans. Finally, the jury filed in, pronounced the verdict of "Guilty," and filed out, the judge heaved a sigh of relief and married us then and there.

We flew to San Francisco for a 2-day honeymoon. This was only the *hors d'oeuvre*. The real honeymoon shall take us to Mexico City in December. San Francisco was as beautiful as ever.

The most anxious moment of this week was 2 nights before the wedding when Imme and I went to a movie. Imme was trying on her ring and dropped it on the floor into a ventilator. Like the perfect gentleman I am, I crawled down outside the building and along about 100 feet of pitch-black and filthy and extremely narrow passages while Imme hammered on the floor above to guide me. I actually found the ring and got out alive, much to my surprise. I believe in the Middle Ages they had similar ordeals for young men who wanted to get married.

It is worth noting that in his Cambridge University days, Freeman had done some night climbing of buildings on the campus—an old tradition.

On 6 February 1959, I received a letter from Freeman that contained the following paragraphs:

Last Sunday the Mexican Federal Police raided the Club Deportivo Panamericana [a place in Baja California we had visited] collected 63 Americans, confiscated all their cash, and left them to languish in the Tijuana town jail until they could raise the $1600 bail each. The San Diego paper has been keeping this outrage on page 1 all the week. I whistle softly to myself when I think of Imme and Margot sharing a cell on one side of the corridor, you and I on the other. There, but for the grace of God.

By the way, Margot left us last week and found herself a job with a family in La Jolla who have 5 children under 10. So now she is finding out what real Americans are like.

Freeman and Imme had four daughters. On September 1970, Freeman wrote to me about an exchange he had with his then seven-year-old daughter Mia (then Miriam). The late Arthur Wightman mentioned in it was a professor of physics at Princeton and was noted for his work on the mathematics of the quantum theory of fields. Freeman writes,

> This morning I had an illuminating dialog with my 7-year-old daughter Miriam. At lunch yesterday Arthur Wightman scribbled some equations on a paper napkin and I took it home. So the following conversation occurred.
> Miriam. What is this?
> Me. It's a problem which Wightman gave me.
> Miriam. Is it hard?
> Me. It was too hard for Wightman so probably it is too hard for me.
> Miriam. But Daddy, you know that negroes are just as smart as white people.

In 1986, I published an article in the *New Yorker* about climbing Popocatapetl in Mexico.[3] On 18 December, Freeman wrote,

> We have been enjoying your "Breaking in at the *New Yorker*" and even more your Popocatapetl. We have vivid memories of our honeymoon in May 1959 when Imme, then 5 months pregnant, drove us in a dilapidated rental car up from Amecameca to the shoulder between Popo and Inxti. [Iztaccihuatl is the slightly lower twin volcano.] In those days the road was unpaved and there were no climbers. Not even any tourists. The sole human being up on the pass was a young Indian girl collecting bundles of tall grass reeds, presumably to be taken down to the village and woven into hats and baskets. Imme spoke to the woman in Spanish but she did not answer. We wondered how the hell she got up there, and how the hell she would get down again. Probably by walking both ways. And we were feeling quite proud of our bravery getting up and down by car. Leaving her alone up there, when we drove down in the evening, we did not feel like heroes anymore.
> What I liked about your story was the brevity of the climb itself. In the story, as in real life, it is the preparations that are memorable.
> Happy New Year!
>
> Imme and Freeman

Freeman was away from the Institute during some of spring 1958. He was working on the spaceship. So he missed the mystery. Threatening notices were being placed on the doors of the members. One might have dismissed this as a joke, except that Robert Oppenheimer, our director, had lost his clearance not too long before and had been accused of being a Communist and possibly a spy. After Freeman returned, I told him about what had happened. His first response was to accuse me. I said that this was not possible, because the handwriting on the notes was too good. Then he said, without explanation, "Gaudy night." It was only a few years later that I understood the reference. In 1935, Dorothy L. Sayers published a Lord Peter Wimsey crime novel with that title. It deals with the kind of note leaving at Oxford, but of course there is murder involved.[4] Once I saw the connection, I read the novel but was surprised by its casual anti-Semitism. A young man is asked if he is related to Lord Peter. "Why of course," said the young man sitting up on his heels. "He's my uncle; and a dashed more accommodating than the Jewish kind," he added, as though struck by a melancholy association of ideas. No explanation was offered for this bizarre association, which seems in your face (defiantly confrontational; blatantly aggressive or provocative). I wrote a note to Freeman from Oxford where I was spending a year, pointing this out. On 3 February 1972 he replied:

Dear Jerry

Thanks for your note about Gaudy Night. I had totally forgotten that. I have been hoping some time to see "Le Chagrin et la Pitie" but it didn't come here yet. I am now reading "The Double-Cross System." It is the best of the WW2 spy books. The Israelis now have a superb opportunity to make use of its technique to deal the Soviet intelligence apparatus a mortal blow. I hope they are making the most of it.

Whatever it was, the intelligence apparatus seems to have survived. The mystery of the notes at the Institute was solved, when the police, and possibly the FBI, began interviewing people. The young son of one of my physics colleagues confessed. He said it had been a "joke."

The next rather long letter requires some explanation. It is dated 15 December 1971. The first part refers to the fact that by 1971, I had made a few trekking trips to Nepal and had seen most of the great mountains there. In 1969, I had driven to Pakistan and had seen many of the mountains there. The rest of the letter refers to the fact that I was in the midst of writing a profile of Albert Einstein, which eventually appeared in the *New Yorker*.[5] I had mentioned that I was doing this to Freeman.

Like everyone who writes about Einstein, I began my research with his "Autobiographical Notes," first published in 1949.[6] In them, he tells us that at age sixteen, he had hit on a "paradox" that ten years later led him to the theory of relativity. He imagined that he could move as fast as a light wave. He could then attach himself to a crest or trough, and the wave would no longer appear wave-like. Hence he would have a way of determining his speed in an absolute sense. This he felt was impossible, so something had to be wrong. It turns out that relativity forbids motions of a massive object at the speed of light. Einstein does not put things quite this way, but that is the idea. What Dyson pointed out to me was that at age fifteen, Einstein actually wrote a little paper about this and that in this little paper there was no hint of relativity at all. Here is the letter:

Dear Jeremy

Thank you for letting us see the Einstein chapter which Helen gave us yesterday. [This is a reference to Helen Dukas, who had been Einstein's secretary in Germany and emigrated with him. She, along with Einstein's sister, shared his house in Princeton.] I returned it to you, but I would be glad to have a copy to keep when you have one available. Also thanks for the Nanga Parbat picture which is superb. [Nanga Parbat is a 26,600-foot tall mountain in Pakistan.] I have three Nanga Parbat books, by Dyhrenfurth, Herrligkoffer and Buhl [who first climbed the mountain] and the whole thing has for me a mythical quality. The fact that you have looked down [from a plane] on that Olympian summit plateau makes you almost a mythological character too.

To return to Einstein. Recently I read the article "Über die Untersuchung des Aetherzustundes im magnetischen felde" ["On the investigation of the aether state in a magnetic field"] which Einstein wrote at the age of 15 and I wonder if you have read it. It appeared finally in "Physikalische Blätter" Vol. 27, 385 (1971). I found it extremely illuminating. In fact, it has very much the same value for the understanding of Einstein's development as the decipherment of Linear B had for the understanding of Greek history. In both cases, we were suddenly and unexpectedly confronted with a written document from a period far earlier than anyone believed possible. In both cases, the publication of the documents was attended with enormous complications, with Sir Arthur Evans and Otto Nathan playing possibly analogous roles.

My own reaction to Linear B and to the Einstein article went through the same three stages. First, with great enthusiasm and hope that some great mystery would be revealed.

Second, deep disappointment that the documents turned out to be so completely ordinary. Thirdly, understanding that precisely ordinariness of the documents was the mystery, making the miracles that came later even more miraculous.

When one looks in detail at the Einstein article, a number of things become immediately clear. His main concern was to find some experimental way of checking whether the mechanical modes of the aether were true. He was obviously impressed by the elegance of Hertz's experiments and the contrast between the elegance of the experiments and the clumsiness of the mechanical models. But he never explicitly questions the mechanical picture.

The most remarkable thing about the article is that it contains not the slightest hint of relativity. Not a hint of the famous question of what you see when you travel along with a light-wave. But it does contain two things that were decisive for the later history. (1) For Einstein the electromagnetic field came first, before he had been exposed to the whole imposing apparatus of Newtonian mechanics. So when he found out that Maxwell and Newton were in contradiction it was natural for him to hang on to Maxwell and discard Newton. The orthodox nineteenth-century education had of course precisely the opposite emphasis. (2) Einstein at 15 was not yet thinking in terms of a Gedanken experiment. He was discussing what he thought of as a real experiment. But the kind of experiment he proposed would have naturally led him to consider others which would be genuinely Gedanken in character. The style of his thinking is already going in the direction which would lead him to 1905.

I hope you will in your account of Einstein give some weight to the 1895 article. I think it is important for putting his work in perspective, just as Linear B has been for Greek history.

For the last two months I have been in the middle of great personal dramas concerning Einstein. How Einstein would laugh if he could see the learned men squabbling over his relics. I would laugh too except that the drama is a tragic one. Meanwhile I have been reading over again Einstein's great papers, the 1905 and the 1915 and finding them as fresh and wonderful as ever.

I hope you have a splendid time in Oxford. I probably won't see you before you go. But let's keep in touch.

Yours ever,
Freeman

The standard view for the propagation of light in the nineteenth and early twentieth centuries was that it propagated in a mechanical medium, which was known as "ether," just as sound waves propagate in a medium.

What Freeman was noting was that at age fifteen, Einstein accepted this and even proposed an experiment to detect the ether. In this paper, there was no hint of what he would later write in his relativity paper: "The introduction of a 'luminiferous ether' will prove to be superfluous."[7] The "squabble" Freeman refers to had to do with the trustees of the Einstein estate blocking the publication of his letters. I had such an encounter with one of them, Otto Nathan. He held up the publication of my *New Yorker* profile of Einstein by making all sorts of demands before he would allow us to use quotations. At one point, he told me that the estate had the rights to the formula $E = mc^2$. I finally had to pay a good deal for these rights. Soon after the first part of the profile had been published, Nathan called me to complain that he had not been credited.

Here is a letter dated 14 October 1968, just after I had taken a job at the Stevens Institute of Technology:

Dear Jerry

I was considerably moved by your letter. I had a very good feeling about Stevens when I went there for a day last winter, and I hoped you would see it the same way. It seems you did. . . . I knew you were going through a rough time. Everything in your letter increases my respect for Stevens and for you.

The last year was also a restless one for me. . . . I did this year teaching at Yeshiva and I was seriously considering taking a permanent position there. I had come to one of those middle-age crises, feeling maybe I am too old to produce ideas anymore and I ought to settle down to a teaching job before it is too obvious.

Anyhow I finished the year at Yeshiva and came back here. And I have been in the full tide of happiness these last days, proving that one-dimensional Ising models with an interaction going like *distance*$^{-\alpha}$ have a phase transition [as conjectured by Mark Kac] when $1 < \alpha < 2$. Now it is midnight and I just wrote the last words of the proof in a strangely luminous state of mind, as Gibbon describes in his autobiography how he wrote the last words of the "Decline and Fall."

So I am not too old after all, and still have something to do here. As you say, it is nice to know that given the choice there really is no choice.

I would much enjoy another visit to Stevens now you are there. The Nepal pictures are still to be seen, *2001* to be discussed, and much else. Your piece about nuclear weapons is good. It is difficult to write long sentences with clarity, but you did.

Love from us all,
Freeman

The Ising model is a mathematical model that describes how certain solids can become magnetic at critical temperatures. The simplest such model would be a string of tiny magnets that can point up or down. If at some critical temperature all the tiny magnets line up, then the string becomes magnetic. What Freeman showed was that, given the conditions of his proof, there must be such a critical temperature for a one-dimensional string. It was a tour de force bit of mathematics.[8]

This final letter is dated 6 May 1969 and was written from the faculty club of the University of California at Santa Barbara:

Dear Jeremy

This evening Shawn [William Shawn, the editor of the *New Yorker*] telephoned and said he will print my stuff.[9] Naturally I was pleased. I write now to thank you for your services as a midwife. Also thank you for the invitation to talk at Stevens without which this would never have happened.

I promised to send Arthur Clarke a copy of the lecture on space travel, but I don't have his address. If you happen to have it, please send it on a postcard.

By a strange coincidence, so soon after writing this piece for the *New Yorker*, I had again a close encounter with violent death. It came like the Hiroshima bomb on a peaceful sunny day. I was woken by a shattering explosion, followed by moans and cries for help. I was too scared and stunned to run down immediately. While I hesitated a man was burning to death. By the time I came down he had already dragged himself into the children's pool below my window and put out the flames. If I had come down sooner, I might well have been able to save his life. He died in hospital two days later.

The bomb was presumably intended only to burn down the Faculty Club where I am staying. Unfortunately, the caretaker found it and it blew up in his hands.

This campus is now very quiet, trying to digest what has happened. The blood and ashes around the pool have been washed away and the children are again splashing in it. And I am once again a survivor with a bad conscience.

All the best,
Freeman

To this day, the murder of the custodian Dover O. Sharp remains unsolved.[10]

NOTES

This essay is reproduced with permission from Jeremy Bernstein, *A Bouquet of Dyson and Other Reflections on Science and Scientists* (Singapore: World Scientific, 2018), 3–16.

1. The original report was Jeremy Bernstein and Freeman Dyson, "The continuous opacity and equations of state of light elements at low densities," report GA-848 (General Atomic division of General Dynamics Corporation), dated 13 July 1959. The main results were later published as Jeremy Bernstein and Freeman Dyson, "Opacity bounds," *Publications of the Astronomical Society of the Pacific* 115 (2003): 1383–1387.

2. Jeremy Bernstein, "Routes," *New Yorker* (24 August 1981): 98–100.

3. Jeremy Bernstein, "Popocatépetl," *New Yorker* (15 December 1986): 116–123.

4. Dorothy L. Sayers, *Gaudy Night* (London: Gollancz, 1935).

5. Jeremy Bernstein, "The secrets of the Old One," part 1, *New Yorker* (10 March 1973): 44–101. See also Bernstein, *Einstein* (New York: Viking, 1973).

6. Albert Einstein, "Autobiographical notes," in *Albert Einstein: Philosopher-Scientist*, ed. Paul A. Schilpp (Evanston, IL: Library of Living Philosophers, 1949), 2–94. See also Hanoch Gutfreund and Jürgen Renn, *Einstein on Einstein: Autobiographical and Scientific Reflections* (Princeton: Princeton University Press, 2020).

7. See the English translation of Einstein's 1905 article, "On the electrodynamics of moving bodies," translated and reprinted in John Stachel, ed., *Einstein's Miraculous Year: Five Papers That Changed the Face of Physics* (Princeton: Princeton University Press, 1998), 123–160, on 124.

8. Freeman Dyson, "Existence of a phase-transition in a one-dimensional Ising ferromagnet," *Communications in Mathematical Physics* 12 (1969): 91–107.

9. Freeman Dyson, "Reflections: Disturbing the universe," published in three parts: *New Yorker* (6 August 1979): 37–63, *New Yorker* (13 August 1979): 64–88, and *New Yorker* (20 August 1979): 36–80.

10. See Frances Woo, "Faculty Club bombing," Associated Students of UCSB Living History Project (4 December 2019), https://livinghistory.as.ucsb.edu/2019 /12/04/faculty-club-bombing. Years after the bombing, physicist and Nobel laureate Robert Schrieffer—who had joined the faculty at the University of California, Santa Barbara in 1980—suggested that the bomb might actually have been intended to harm Dyson himself, as part of a Vietnam War–era protest against nuclear-related research. See Kenneth Klein, "Custodian killed by bomb in 1969 remembered," *University of California, Santa Barbara Daily News* (11 April 1991): 1, 10, https://www.alexandria.ucsb.edu/downloads/m613mz836.

CODA: NOT THE END

ESTHER DYSON

Well, there you have it! But what was it like to be a child of this amazing man? He died at ninety-six, yet somehow he retained his own childhood curiosity and openness to the end, as this book illustrates. Some scientists talk only with other scientists who can appreciate their importance; Freeman regularly engaged with waiters, airline pilots—and both his children's friends and his friends' children. Indeed, he was the same as a parent as he was as a scientist: curious, open to evidence, and willing to follow wherever that led. Boundaries and rules were merely suggestions—easily crossed for the right reasons. He loved us all, but I think that for him, we were also unique experiments waiting to unfold.

Of course, my childhood felt completely normal. Imme and Daddy, the Institute, supper at home with the whole family every night: we were privileged, but we felt ordinary. We had no TV, true, but we had books galore. (Much later, at age fourteen, when I took the paid job I had long aspired to as a clerk in the Princeton Public Library, I finally entered the real world and made an astonishing discovery: most people did not get to take three months off for summer vacation.)

As I grew older, I could see the effect of Freeman's own British upbringing peeping through. He wanted to save us kids from the formality and rigidity that had shaped his own upbringing: being sent away to school where he endured teasing and tormenting, pointless rules and silly constraints. George and I thought his world revolved around us, and in so many ways it did.

The Institute itself—that collection of buildings centered around Fuld Hall—was just half a mile from our house, slightly downhill. Every weekday after breakfast, he kissed us goodbye and headed off to work, coming home in the evening in time for supper around six o'clock. Sometimes in the winter, when it had snowed, he would take our sled to work. We liked that, because it meant we could go visit him at his office after school to retrieve the sled for ourselves. His office was a wonderful place, with a desk, a papier-mâché penguin I had made for him—and the blackboard. It was always filled with words and equations, mostly unintelligible, but sometimes we could erase them and write our own messages.

At supper, he wouldn't quiz us about physics or math; we talked about other kids and our teachers, or perhaps the latest (sanitized) Institute gossip. When he occasionally appeared on television, we would go to the neighbors' house to watch, but home was the center of our universe, we thought, and of his.

In general, he and Imme enabled us rather than directed us. As described so ably in chapter 1, he was not a fan of formal schooling. He never taught us much, but he asked a lot of questions and encouraged *us* to ask questions.

The year we lived in La Jolla (1958–1959), which ended with the birth of my sister Dorothy, was especially magical. We had a fancy house courtesy of General Atomic, with a swimming pool and an orchard—oranges and avocados—out back, and beyond that, a dirt road between our property and a stretch of half-cliff/half-landslide probably thirty feet high/deep. There was also a working faucet. George and I would spray ourselves, slather ourselves in mud (which we called armor), fight until our armor fell off, slide down the landslide, clamber back up, and just have a great time. No math puzzles, no toys. Just a space for imagination and creativity. (And yes, fortunately, there was an outdoor shower near the pool.)

I loved that year in La Jolla, and I loved traveling in general. We had been to both Pontresina (in Switzerland) and Pasadena (NASA's Jet Propulsion Lab), and I couldn't quite remember which was which. As I approached the age of thirteen, I kept hearing about exchange programs, where two students from different locations would trade places for a year. Freeman mentioned my ideal match: Sarah Carthy, the daughter of his Bomber Command/Operations Research friends Dorothy and John Carthy. I jumped on the idea, but Sarah was too busy getting ready for her A-levels the following year to come to the United States, so instead I simply joined the family in London for a year of British schooling. I had heard from Freeman about life in England; I wanted to see it for myself. And from his side, Freeman knew that learning doesn't happen in the classroom as much as outdoors, with another family, in the absolute Elsewhere.

Among other things during that year in London (1964–1965), I spent more time with my aunt, Freeman's sister Alice, who had been something of a good girl who stayed close to the family while Freeman traveled widely and ultimately moved to the United States. Freeman's life was one of exploration while hers was one of service. I sensed he wanted to give us that same freedom.

Meanwhile, as I joined the tenth grade back at Princeton High School, there was the question of language classes. I already knew German from home and French from grade school, so I made another Freeman choice: Russian. He was no fan of the Russian government (or of the US bureaucracy), but he loved the Russian people. His original plan, he told me, was to move to Russia after the war, because that's where so many of the great mathematicians with whom he studied came from. But a little more research—and some job offers—changed that plan, and he ended up in the United States instead. Fortunately!

Anyway, in 1965, I signed up for first-year Russian and followed that with a summer at the Russian-only Institute of Critical Languages in Putney, Vermont, and then joined the three-student fourth-year Russian class the next year at Princeton High. My career goal was to be the Moscow bureau chief for the *New York Times*, combining asking questions with the absolute Elsewhere of Russia. Over the years, I have spent a lot of time in Russia—and learning so much about the United States by looking at it from outside.

Overall, being one of Freeman's children was an amazing experience. I had visions of running away—it just seemed so exciting—but I could never really come up with any excuse for doing so. There wasn't actually anything I wanted to do that my parents would not allow.

Instead, in my junior year of high school, I idly applied to two colleges for early admission and ended up getting accepted to Harvard. Freeman and Imme gave their blessing. Freeman said in truthful jest, "With so many kids [five at that point, and one more on the way] we could definitely use the extra bedroom!"

So off I went to Harvard and managed to pass as a teenager. In fact, I had worked hard at school with the goal of getting into college, but once *in* college, I ended up facing that same education/stifling conundrum that troubled Freeman. I joined the *Harvard Crimson* student newspaper and spent most of my time there, learning to proofread and ask questions and discovering new information rather than regurgitating textbooks. I rarely went to class, and in the spring, the college elders told me that I was welcome to take some time off and come back when I was ready. I decided to hitchhike through Europe and spend the fall with my boyfriend, just-graduated senior Tim Crouse, who was heading to Oujda, Morocco, to serve in the Peace Corps.

Of course, the Peace Corps did not allow stray girlfriends. But at least my parents were resigned to the idea of freedom with responsibility. Perhaps Freeman was remembering his own excursion at the age of seventeen, after his first few months in Cambridge, when he went climbing in Wales and ended up injured in hospital for several days but survived intact. In essence, Freeman told me, "If that's what you want to do, go ahead! But if you get stuck in jail, we won't get you out. If you get pregnant, we won't take care of the baby. But we hope you learn a lot!"

And indeed I did. I hitchhiked through Europe and showed up in a bus station to meet Imme's sister Sigrid in Sinsheim, Germany, pretending that I had arrived by bus. My plans to get to Oujda were so vague that she insisted on buying me a plane ticket (which I learned much later was actually funded by Freeman and Imme).

A few more anecdotes from a long life reflecting this extraordinary influence. In my mid-twenties, while I was working at a Wall Street firm, I was mugged on the street. The next day (a Saturday), Freeman and Imme came to take me to lunch at Rockefeller Center. I don't remember much of that, but I do remember Freeman running up the down escalator in some attempt to shake off the excess of emotion.

A few years after that (1982), I took over a newsletter and conference business focused on the tech sector, PC Forum—originally Personal Computer Forum, and ultimately (when I sold it in 2005) Platforms for Communication Forum. While the other kids had weddings and horse shows, I had this conference, and it became an annual family event. (We encouraged all the attendees to bring their families; it made everyone behave better, and even

Freeman Dyson (center, holding granddaughter Lauren) talks with another participant at the annual PC Forum meeting in Tucson, Arizona, 1990. Photograph by Ann E. Yow-Dyson, Getty Images.

Freeman and Imme Dyson at the Baikonur Cosmodrome in Kazakhstan in March 2009 to watch the launch of Soyuz TMA-14 Spacecraft Mission, for which Esther Dyson trained as a backup crew member. Photograph by George Dyson.

in those days, tech people tended to neglect their families.) Freeman and Imme and George came regularly; the sisters and their families often came too, but not quite every year.

Freeman loved it. Perhaps 10 or 20 percent of the attendees knew who he was, and some smaller number were excited to talk with him, which was just about the right proportion. In turn, he loved talking with *them*, whether it was Bill Gates or some start-up visionary with a crazy idea for an online community. And he loved hanging out with his grandchildren by the pool too.

And then in 2008 there was another crazy idea. As you know by now, Freeman was a wonderful parent: Do what you like; just take responsibility for the consequences. This time, there was another lunch in New York, and I announced plans to spend half a year in Russia, training as a backup cosmonaut for a trip to the International Space Station. (I could afford the training but not the actual trip. Coincidentally, the person I was backing up was Charles Simonyi, author of Microsoft Word and a long-time member of the Institute's board.) I would learn about space plumbing and space medicine, spend time with both Russian cosmonauts and NASA astronauts, and experience bygone Soviet Russia in a Russian government facility: the Yuri Gagarin Cosmonaut Training Center.

This time there was no measured "Whatever you like, dear." Instead, Freeman jumped up and hugged me with joy. Suddenly I understood that I was his first-born child and that I was about to take that next step toward his lifelong dream of going into space.

Some nine or ten months later, he and Imme and George came out to Baikonur, Kazakhstan, to join me to see the launch of the Soyuz TMA-14 Spacecraft Mission for which I had trained. Ironically, for all of Freeman's connections with the US space program and frequent invitations to launches, there were always glitches and delays and he had never seen an actual crewed mission launch to space from close up. I'm so glad I was able to give him that pleasure and a glimpse of the future that is still ahead.

ACKNOWLEDGMENTS

This project began at the suggestion of Jermey Matthews at the MIT Press. Soon after Freeman Dyson passed away, Jermey floated the idea of developing a book aimed at nonspecialists, in which authors could explore facets of Dyson's extraordinary career. I was immediately hooked by the idea, and I remain grateful for Jermey's enthusiastic support of this project at every stage. My thanks to Haley Biermann at the MIT Press as well for her logistical help and to Jermey and Haley for securing constructive and thorough referee reports from three anonymous reviewers.

I am also grateful to Mark Wolverton, who served as the developmental editor for the book, working with each chapter author to sharpen drafts and help craft a coherent whole from so many parts. I first met Mark when he held a Knight Science Journalism fellowship at MIT during 2016–2017, and it has been a pleasure to work with him on this book.

It has also been a delight to work with Laurent Taudin, who prepared original illustrations for this book. He read the entire manuscript with gusto, shared suggestions on the text, and delved into the illustrations with creative flair.

My understanding of Freeman Dyson's life and legacy has been enormously enriched by conversations with each of the chapter authors who contributed to this book; they all have my sincere thanks. Among these scholars, colleagues, and friends, George Dyson stands out for the generosity with which he has shared his encyclopedic knowledge of relevant dates and events in his father's life. George also supplied myriad documents and photographs to several chapter authors and patiently fielded our many questions.

I have benefited from discussions about Dyson's work with William Press, who not only shared his insights but also generously shared a digitized corpus of much of Dyson's work that Press had prepared while working with Ann Finkbeiner on the official biographical memoir about Dyson for the US National Academy of Sciences. This book was conceptualized and completed entirely amid the disruptions of the global COVID-19 pandemic, which—among so many other impacts—limited authors' access to physical libraries and related resources. The digital materials that Press shared made it possible to complete this book.

My introductory essay for this book benefited from suggestions from Clare Sestanovich, Michael Gordin, George Dyson, and Mark Wolverton. And my work on the history of quantum electrodynamics was incomparably enriched by Freeman Dyson himself, who generously answered my questions during a long interview back in 2001 and—in a gesture that still astonishes me—shared with me his trove of precious family letters.

Jeremy Bernstein completed his PhD in physics at Harvard University in 1955, studying under Nobel laureate Julian Schwinger. He served on the faculty at New York University and the Stevens Institute of Technology and has held appointments at Brookhaven National Laboratory, CERN, Oxford, the University of Islamabad, and the Ecole Polytechnique. A prolific author, Bernstein was a staff writer at the *New Yorker* magazine between 1961 and 1995 and has published twenty-nine books, including *Einstein* (1973), *Hans Bethe: Prophet of Energy* (1980), *The Life It Brings: One Physicist's Beginnings* (1987), *Quantum Profiles* (1990), *Hitler's Uranium Club: The Secret Recordings of Farm Hall* (1996), *Oppenheimer: Portrait of an Enigma* (2004), and, most recently, *A Bouquet of Dyson and Other Reflections on Science and Scientists* (2018).

Robbert Dijkgraaf is Minister of Education, Culture, and Science for the Netherlands. Between 2012 and 2022 he served as director and Leon Levy Professor of the Institute for Advanced Study in Princeton, and as a distinguished university professor at the University of Amsterdam. A mathematical physicist, he has made important contributions to the study of string theory and black holes. He is a distinguished public policy adviser and passionate advocate for science and the arts, having served as president of the Royal Netherlands Academy of Arts and Sciences and the InterAcademy Partnership, the global alliance of science academies advising the United Nations. For his contributions to science, Dijkgraaf was awarded the Spinoza Prize, the highest scientific award in the

Netherlands. He holds honorary doctorates from Leiden, Brussels, and Nijmegen; is a member of the American Academy of Arts and Sciences and the American Philosophical Society; and was named a Knight of the Order of the Netherlands Lion. Many of his activities—which include frequent appearances on Dutch television, a monthly newspaper column, and the launch of the science education website, Proefjes.nl—are at the interface between science and society. He is the author of *The Usefulness of Useless Knowledge* (2017), articulating how basic research is essential to innovation and societal progress.

Esther Dyson was Freeman's first biological daughter, half-sister to Katarina, and born in 1951. In some ways, she got the best of him as a father: almost full attention and family dinner almost every night. As the family grew bigger and he traveled more, she followed his path and left home early—moving into the business world as a journalist and Wall Street analyst in the nascent market of the personal computer and the Internet. Yet she kept in touch; while her siblings got married and held weddings, she invited the family to her annual PC Forum conference for the tech elite. Freeman and Imme were faithful attendees. In more formal terms, Esther became a leading analyst of the emerging IT marketplace and then left that world to focus on health and social equity as executive founder of a ten-year nonprofit project, Wellville (2015–2024). Wellville operates in five small US communities and takes a practical approach to social problems that reflects Freeman's lifelong habit of always asking "Why?" or, if need be, "Why not?"

George Dyson, Freeman's son, is an independent historian of technology whose subjects have included the development (and redevelopment) of the Aleutian kayak (*Baidarka*, 1986), the evolution of artificial intelligence (*Darwin among the Machines*, 1997), a path not taken into space (*Project Orion*, 2002), the origins of the digital universe (*Turing's Cathedral*, 2012), and why analog computing is destined to regain control (*Analogia*, 2020).

Ann Finkbeiner is a freelance science writer who has written regularly for *Nature, Science, Hakai,* and *Scientific American* about astronomy, cosmology, and grief, among other subjects. She also writes about the science of national security, specifically about Jason, the half-century-old group of independent academic scientists who advise the US government's executive branch. Her books include *A Grand and Bold Thing* (2010) about the Sloan Digital Sky Survey and *The Jasons: The Secret History of Science's Postwar Elite* (2006). She is the co-proprietor of a group science blog, "The Last Word on Nothing."

Amanda Gefter is an award-winning science journalist who writes about physics, cosmology, cognitive science, and philosophy. She is the author of *Trespassing on Einstein's Lawn* (2014) and cohost of the BookLab podcast. She lives in Watertown, Massachusetts.

Ashutosh Jogalekar trained as a chemist and works on building intelligent models for molecular design. A desire to communicate science to a broader audience led him to moonlight as a science writer. In this capacity, Ash has written for *Scientific American*, *Nature Chemistry*, and the Lindau Nobel Laureate Meetings. He is currently a columnist for the popular website 3 Quarks Daily. An email to Freeman Dyson about an obituary of Dyson's mentor, physicist Hans Bethe, which Ash had written for *Physics Today*, led to a decade of memorable friendship and mentorship. He lives near San Francisco with his wife, a computer scientist, and their one-year-old daughter whose mind he regards as the best model for understanding how to build intelligent models.

David Kaiser is Germeshausen Professor of the History of Science and professor of physics at MIT. He is the author of several award-winning books about modern physics, including *Drawing Theories Apart: The Dispersion of Feynman Diagrams in Postwar Physics* (2005), which received the Pfizer Prize from the History of Science Society for best book in the field; *How the Hippies Saved Physics: Science, Counterculture, and the Quantum Revival* (2011), which received the Davis Prize from the History of Science Society for best book aimed at a general audience and was named Book of the Year by *Physics World* magazine; and *Quantum Legacies: Dispatches* *from an Uncertain World* (2020), which was honored as among the best books of the year by *Physics Today* and *Physics World* magazines. A fellow of the American Physical Society, Kaiser has received MIT's highest awards for excellence in teaching. His work has been featured in *Science*, *Nature*, the *New York Times*, and the *New Yorker*. His group's recent efforts to conduct a "Cosmic Bell" test of quantum entanglement were featured in the documentary film *Einstein's Quantum Riddle* (2019).

Caleb Scharf received his doctorate in astronomy from the University of Cambridge and has a research career that ranges from observational cosmology and astrophysics to extensive work in exoplanetary science and astrobiology. He is director of astrobiology at Columbia University and codirector of Columbia's Habitable Planet Network, and has served as a global science coordinator for the Earth Life Science Institute at Tokyo Tech, and cofounder of the nonprofit science institute YHouse in New York City and Princeton. In addition to research and teaching, he is a prolific science writer who has authored a prize-winning astrobiology textbook, four popular science books, and articles for many publications, including more than five hundred pieces for *Scientific American*.

William Thomas is a science policy analyst at the American Institute of Physics. He writes and edits for the FYI science policy news service (www .aip.org/fyi). He writes particularly often on science programs at the Department of Energy, NASA, and the Defense Department, as well as on science and technology policy deliberations in Congress. He earned his PhD in the history of science from Harvard University in 2007 and is the author of *Rational Action: The Sciences of Policy in Britain and America, 1940–1960* (2015), which focuses on the early history of operations research (OR). He has previously worked as a historian at the American Institute of Physics Center for History of Physics and Imperial College London, and, in addition to OR, has also written on subjects such as the early history of particle detection and the history of research on the West Antarctic Ice Sheet.

INDEX